SpringerBriefs in Applied Sciences and Technology

SpringerBriefs present concise summaries of cutting-edge research and practical applications across a wide spectrum of fields. Featuring compact volumes of 50 to 125 pages, the series covers a range of content from professional to academic.

Typical publications can be:

- A timely report of state-of-the art methods
- An introduction to or a manual for the application of mathematical or computer techniques
- A bridge between new research results, as published in journal articles
- A snapshot of a hot or emerging topic
- An in-depth case study
- A presentation of core concepts that students must understand in order to make independent contributions

SpringerBriefs are characterized by fast, global electronic dissemination, standard publishing contracts, standardized manuscript preparation and formatting guidelines, and expedited production schedules.

On the one hand, **SpringerBriefs in Applied Sciences and Technology** are devoted to the publication of fundamentals and applications within the different classical engineering disciplines as well as in interdisciplinary fields that recently emerged between these areas. On the other hand, as the boundary separating fundamental research and applied technology is more and more dissolving, this series is particularly open to trans-disciplinary topics between fundamental science and engineering.

Indexed by EI-Compendex, SCOPUS and Springerlink.

Marcus Vinicius da Silva Neves ·
Antonio Felipe Flutt

Energy Efficiency in Oil Production

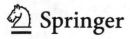

Marcus Vinicius da Silva Neves
Federal University of Rio de Janeiro
Rio de Janeiro, Brazil

Antonio Felipe Flutt
Petrobras University
Rio de Janeiro, Brazil

ISSN 2191-530X   ISSN 2191-5318 (electronic)
SpringerBriefs in Applied Sciences and Technology
ISBN 978-3-031-54273-2   ISBN 978-3-031-54274-9 (eBook)
https://doi.org/10.1007/978-3-031-54274-9

This Springer imprint is published by the registered company Springer Nature Switzerland AG
The registered company address is: Gewerbestrasse 11, 6330 Cham, Switzerland

Paper in this product is recyclable.

Foreword

I received with great pride and a deep feeling of purpose the invitation to introduce the book *Energy Efficiency on Oil Production*. Written by Marcus Vinicius da Silva Neves and Antonio Felipe Flutt, this book represents a source of firm knowledge in a world that is juggling the demands of both deep decarbonization and a fair energy transition.

Fundamentally, *Energy Efficiency on Oil Production* is an heir to the writers' unwavering commitment, in-depth training, and wealth of expertise. With an illustrious academic and professional journey, mechanical engineers Flutt and Neves amass invaluable contributions to PETROBRAS and academia, along with their backgrounds in thermodynamics and mechanical engineering, that provide the book a rare and priceless practical and theoretical richness.

The relevance of this work cannot be diminished, especially in light of the present energy revolution. With the globe looking for sustainable solutions to move toward a low-carbon future, the knowledge offered in this book is essential. It places energy efficiency in oil production in a larger context of environmental stewardship and sustainable development in addition to addressing its technical components.

With real-world examples, the book provides a solid foundation for solving practical oil production issues. It closes the knowledge gap between theory and practice by outlining a number of best practices that are crucial for the operation of oil-producing units and by presenting contemporary techniques for energy efficiency analysis.

Having my fair share of experience in the O&G sector, I anticipate this book becoming a cornerstone for professionals, analysts, students, and designers who are interested in energy efficiency because of its educational approach. An introduction to thermodynamics and energy efficiency is presented at the outset, providing a strong basis for comprehension of the ideas that follow. Notable research has been done on the exergy concept, energy return on investment (EROI), and how these factors affect how competitive the oil producing industry is. The importance of energy efficiency in raising the productivity and profitability of oil-producing facilities is highlighted in these sections.

This book stands out because of its focus on using energy efficiency analysis in the oil business. It provides readers with an extensive list of best practices, real-world examples, and up-to-date analytical techniques, going beyond academic discourse. This practical method is intended to enable experts in oil-producing facilities to optimize energy usage, making it an invaluable resource in the current energy industry.

To sum up, "Energy Efficiency on Oil Production" is a noteworthy contribution to academics and industry. It displays the writers' extensive expertise and dedication to the fields of oil production and energy efficiency. This book provides direction, inspiration, and workable answers as we endeavor to achieve a sustainable future while navigating the challenges of the energy transition. It represents a seminal undertaking that will hopefully influence how the oil sector approaches energy efficiency in the future.

I hope the reader will, through this work, see another evidence of the still long-lasting relevance of the Oil and Gas sector. Albeit challenged by the requirements of decarbonization, there still room for astonishment, by admirable feats of wisdom and acumen like the work of Flutt and Neves.

Rio de Janeiro, Brazil   Jean Paul Terra Prates
                                                                                        President of PETROBRAS

Acknowledgements

In the amazing process of writing *Energy Efficiency on Oil Production*, cooperation and support played a crucial role. This book, a synthesis of engineering knowledge and real-world experience, is a tribute to the hard work of many people, for whom we are eternally grateful.

To our distinguished professors at the postgraduate engineering programs of COPPE/UFRJ. Our technological expertise and intellectual development have been largely fueled by their demanding academic preparation and wise mentoring.

To PETROBRAS, more than just a company, a driving force behind our advancement as professionals. Our journey has been made possible by their consistent support and dedication to creating an environment that encourages learning and innovation.

To our parents, our initial mentors and enduring role models. Our road through obstacles and uncertainties has been illuminated by their wisdom, sacrifices, and unwavering faith in our skills.

To our spouses, our sources of wisdom and support. Their endurance, forbearance, and unwavering assistance in striking a balance between the demands of this endeavor and their familial obligations have been nothing short of remarkable.

And the brightest stars in our cosmos, our children. Their boundless curiosity, upbeat dispositions, and pure excitement have served as a continual source of inspiration and a constant reminder of the better future we are working to create.

To our students, friends, and colleagues in the area, whose opinions and insights have made our work far better. Their viewpoints have expanded our horizons and enhanced our comprehension.

And lastly, to the community of experts and engineers working in the energy industry. We are continually inspired by your commitment to sustainable methods and to developing the discipline, which has greatly influenced the content of this book.

Marcus Vinicius da Silva Neves
Antonio Felipe Flutt

Contents

About the Authors

Marcus Vinicius da Silva Neves is Equipment Engineer at PETROBRAS since 2002. He has a M.Sc. in mechanical engineering from the Federal University of Rio de Janeiro (COPPE/UFRJ) and is doing his Ph.D. studies in energy planning program at the same university. He has been certified as Energy Manager by the American Society of Energy Engineers (AEE). At PETROBRAS, he has held several managing positions, including Maintenance Engineering Manager, Operational Safety Manager, Offshore Installation Manager, and now he acts as Petrobras University's manager of the center of science and technologies related to petroleum and geosciences. He is author of the book Energy Efficiency in Petroleum Production and in Thermoelectric Power Plants and has actuated as Collaborator Professor at Fluminense Federal University (UFF) and Rio de Janeiro Federal University (UFRJ), among others, for post-graduation programs.

Antonio Felipe Flutt Mechanical Engineer (Federal University of Rio de Janeiro, UFRJ, 1990, Brazil—Academic Dignity at MAGNA CUM LAUDE for the performance shown), postgraduate degree in oil and gas equipment engineering (PETROBRAS, 2003, Brazil), M.Sc. in Mechanical Engineering (Federal University of Rio de Janeiro, UFRJ/COPPE-2008, Brazil). Collaborating professor at the Federal University of Rio de Janeiro (UFRJ), the State University of Rio de Janeiro (UERJ), the Federal Center for Technological Education of Rio de Janeiro (CEFET/RJ) and the Pontifical Catholic University of Rio de Janeiro (PUC-Rio) for postgraduate programs. Based at the Petrobras Corporate University in Rio de Janeiro, he has been employed by PETROBRAS since 2002. He holds the position of Equipment Engineer and has worked as a teacher, senior consultant, and coordinator of training, specialization, and improvement courses in the fields of flow, thermodynamics, and dynamic equipment. He is also the coordinator of the Equipment Engineering Training Course—Mechanics and is in charge of creating programs for engineering selection processes. And now he acts as Petrobras University's general manager. He is also author of the book Energy Efficiency in Petroleum Production and in Thermoelectric Power Plants.

Chapter 1
Introduction—Thermodynamics and Energy Efficiency

Thermodynamics has traditionally been defined as the study of heat and work. The steam engine's invention in the nineteenth century was closely related to Newcomen's discovery, and it developed at the same time as the Industrial Revolution. This succinct historical synopsis highlights the close relationship between science and technology development; it's a great example of how scientific bases encourage and assist engineering. An industrialized, globalized world that is insatiably hungry for energy nonetheless faces many of the same problems, concerns, and paradigms as it did in the past.

Before delving further into thermodynamics, it is important to note that understanding its principles is a prerequisite for understanding how energy is transformed inside a system. The first law of thermodynamics states that energy can only be transformed from one form to another; it cannot be created or destroyed. From a different angle, the second law or the entropy law suggests that energy tends to become more dispersed and less accessible for use.

From that point, its refinement and advancement were propelled by Watt's enhancements to the steam engine. Watt aimed to augment and perfect Newcomen's design to yield a higher quantum of mechanical work from a progressively reduced heat consumption; in other words, to achieve greater output with less coal combustion.

Before delving further into the nuances of energy efficiency, it is critical to underscore the relationship between existing energy sources and their renewable potential or, in other words, the renewable potential of energy conservation processes. Recently, much discourse has revolved around sustainable development, greenhouse gas emissions, environmental impact, and renewable energy sources. Until now, renewability has been associated with mass balances without taking into account the reduction in availability (exergy destruction) associated with conversion processes.

The current paradigm for human development is the use of fossil fuels at a rate higher than the generation of fossil fuel storage. Since complete recycling is not feasible, it is also imperative to look for methods that maximize the energy available from all accessible sources, including so-called renewable sources.

© The Author(s), under exclusive license to Springer Nature Switzerland AG 2024   1
M. V. da Silva Neves and A. F. Flutt, *Energy Efficiency in Oil Production*,
SpringerBriefs in Applied Sciences and Technology,
https://doi.org/10.1007/978-3-031-54274-9_1

With detrimental consequences for the environment and public health, humanity has become increasingly dependent since the Industrial Revolution on non-renewable energy sources like coal, oil, and natural gas to drive economic growth and enhance standards of life. The energy sector's emissions of greenhouse gases, especially carbon dioxide (CO_2), are a major cause of global warming even with the more recent advancements in renewable energy sources like wind, solar, and biomass energy. Between 1900 and 2020, the energy sector's CO_2 emissions increased by more than 300 times, mostly as a result of rising energy demand brought on by urbanization, industrialization, and population growth.

When we apply the laws of thermodynamics to the subject of energy efficiency, we can observe that entropy growth will always result in some energy loss, even in cases where increasing the efficiency of energy transformation is possible. Therefore, it is imperative to look for strategies that can lower these losses, like using more efficient technology and figuring out how to better utilize wasted energy.

Carnot laid the groundwork for the systematic study of thermodynamics and the interplay between thermal and mechanical energies, drawing inspiration from the operation of the energy conversion machinery. The Conservation of Energy principle, asserting the immutable transformation of energy from one form to another without creation or annihilation, mirrored this apparatus. The objective extended beyond mere understanding of physical phenomena, focusing on the development of technology that was not only feasible but efficient. The goal was to engineer a device that not only converted thermal energy into work but did so with minimal resource expenditure.

The notion of work potential and the effectiveness of converting it were not novel; Joule, Clausius, and Lord Kelvin Thomson, among others, had examined it. The second law of thermodynamics was developed as a result of their research. Watt was therefore motivated to advance technology because he wanted to improve the steam engine, which was expected to dominate the market at the time. However, the engine's efficiency was the main cause for concern. The investigation went far beyond defining the machine's bounds. It gave rise to a vibrant new branch of physics with countless applications, from understanding the structure of the universe to explaining the behaviors of gases.

As was previously established, the first law of thermodynamics, or energy conservation, extended the concept of conservation well beyond the domain of steam engines by bridging a new type of energy in motion to mechanical effort. On the other hand, the second law of thermodynamics showed that some processes that seemed to fit inside the first law were fundamentally impractical. It proved the assumption that creating order out of chaos required a large energy expenditure.

The capacity to produce effects or the potential from which all effects generated between interacting materials originate can be used to describe energy itself. It may be identified in two macroscopic forms: potential energy and kinetic energy. It also exists in a form called thermal internal energy that is not apparent at the macroscopic level. However, heat and work are not characteristics of a system; rather, they frequently materialize when a system interacts with its surroundings. This book primarily focuses on the knowledge that energy transfer and conversion play through the lens of thermodynamics, which is crucial to the analysis of energy efficiency. The concept of energy is one of the most essential in the field of thermodynamics because it serves as the basis for the complete study of the subject.

As we continue to analyze energy efficiency, it is a central topic in all discussions about environmental responsibility and reducing greenhouse gas emissions. Energy efficiency, according to author Vaclav Smil of "Energy and Civilization: A History," is the amount of energy used to provide a given good or service. Reducing energy consumption and, consequently, greenhouse gas emissions can be effectively achieved by increasing energy efficiency.

In contrast, energy efficiency by itself may not always be sufficient to achieve sustainability, according to author Benjamin K. Sovacool of "Energy and Ethics: Justice and the Global Energy Challenge." This is because of a phenomena called the "rebound effect," which states that increasing energy efficiency can frequently lead to an increase in energy consumption since the greater efficiency makes energy more readily available and affordable. As a result, it is crucial to investigate other options, such switching to renewable energy sources and reducing overall energy consumption.

Since it's important to remember that improving energy efficiency might not be sufficient on its own, other strategies must be considered. For example, building a grasp of how energy is transformed within a system and finding ways to minimize energy losses require a solid understanding of the laws of thermodynamics. Switching from one energy source to another has both advantages and disadvantages. While it might increase financial burdens and worsen environmental problems, it can also increase people's comfort and well-being.

During the twentieth century, fossil fuels such as coal, oil, and natural gas took the place of more traditional energy sources like firewood and charcoal as the world's main energy sources. Because burning these fossil fuels emits large amounts of greenhouse gasses, the energy matrix has changed, which has accelerated the release of CO_2. At the start of the twenty-first century, CO_2 emissions from the energy sector are still rising, but renewable energy sources are gaining traction as well. Consequently, CO_2 emissions from the energy industry rank among the world's most serious environmental problems at the moment.

Forecasts indicate that over the next 10 years, the share of renewable energy sources in the global energy mix is predicted to increase by 12% on average; however, overall worldwide energy consumption is still rising, particularly in developing countries. Businesses in the energy sector must invest in more efficient solutions and renewable energy sources in order to meet the increasing demand without reducing the standard of living for future generations.

Globally speaking, we have to admit that there is an increasing need to rise energy access for the vast majority of people who are not yet benefiting from it. This need is about fairness and equity in the allocation of energy, not merely efficiency. Worldwide, there is a boost in energy, contingent on local conditions and national energy strategies. The Asia–Pacific region is seen as a leader in the growth of energy due to the rapid economic development of nations such as China and India. Nonetheless, there is a rise in energy consumption in other regions as well, such as Africa and

Latin America. Therefore, even with more effective solutions, mankind still has to provide access to energy for a large number of marginalized people.

The fact that interest in renewable energy sources is expanding is also stressed, demonstrating that the expansion of energy does not always equate to a rise in the use of fossil fuels. As a result, the transition to a more efficient and sustainable energy matrix is becoming a global priority. To reduce greenhouse gas emissions and lessen the effects of climate change, several countries are reevaluating their reliance on fossil fuels and exploring cleaner, renewable alternatives.

It's a challenging and intricate process that involves a number of organizations, including corporations, governments, civil society, and consumers. Money must to be set aside for the research and development of greener and more efficient technologies in order for it to be feasible. It is also necessary to support public policies that improve energy efficiency, decarbonization strategies, and the utilization of renewable energy sources.

The most promising energy sources are those that are recognized as renewable, such as biomass, solar, wind, and hydropower. Particularly in recent years, there has been a significant surge in the usage of solar and wind energy due to falling production costs and incentive policies put in place by various countries. Despite its detrimental effects on the environment and the economy, hydropower is still a major source of renewable energy in many parts of the world.

Therefore, in addition to technical and financial considerations, the relationship between society and energy use is an important issue to consider. How people use energy can have a big impact on how much is consumed and how many natural resources are used. Raising people's awareness of consumption is essential for reducing waste and encouraging the adoption of more environmentally friendly habits. Many questions remain about how energy consumption will develop in the future, including how demand will change over time, how source pricing will change, and whether cleaner, more efficient technologies will become available. It is crucial to investigate these uncertainties while making decisions about future energy and environmental policies.

Thus, the topic that is being provided highlights the significance of using resources in a more sustainable way, which justifies the creation of methods and instruments for assessing industrial facilities' energy and energy efficiency. Therefore, two factors need to be considered when talking about a product's renewability:

- The source of energy.
- Processes used to convert energy efficiently.

Consequently, increasing production and putting as many goods on the market as you can while consuming the least amount of energy is the goal of enhancing energy efficiency in an industrial facility. The financial advantages are clear-cut and immediate. On the other hand, cutting energy use to a minimum directly and proportionately lowers greenhouse gas emissions. Because it has an immediate and permanent effect and eliminates part of the emissions that would otherwise be released into the atmosphere by traditional processes, the use of energy efficiency is therefore configured as the primary method of compensating greenhouse gas emissions. This is in contrast to other traditional methods of compensation, like reforestation.

In summary, the discussion about energy efficiency involves more than just technology developments; it also takes socioeconomic, environmental, and ethical factors into account. In the twenty-first century, the challenge is not only to develop more effective energy solutions but also to make sure that these solutions are equitable and accessible, assisting in the empowerment of marginalized groups and protecting the environment for coming generations.

The adoption of various energy efficiency methods has been identified as one of the primary means of combating climate change caused by greenhouse gas emissions in several projections at the national and even global levels, in light of the ongoing discourse on energy efficiency and its multifaceted implications. Based on a comparative analysis, the first ten years of this century had a greater rate of emissions than growth in energy demand. This was primarily because of a shift in the global energy grid toward a greater proportion of fossil fuels. This percentage ratio may, however, reverse due to mitigating measures involving energy efficiency optimization techniques, albeit with inevitable absolute growth.

Rationalization, which can result in budget cuts, rational resource use—a more involved process that can be in line with sustainability and the use of renewable resources for a new social paradigm—and the redefinition of efficiency concepts in terms of energy and economics are some of these ideas for better managing natural resources. The production of electricity is a prime example of the shift toward increased efficiency. Historically, the world's major electric power plants have produced a significant amount of thermal pollution due to their failure to utilize the residual heat generated (open cycle), as well as atmospheric pollution from burning fuel and radioactive pollution from nuclear unit residue. The exception to this has been hydroelectric power stations. The quest for high efficiency has now become necessary on all scales due to changing circumstances, which have led to the advancement of some technologies and the adoption of more sophisticated ones by others. This has made small- and medium-scale energy generation competitive, especially in more remote areas.

More broadly, as energy is a self-conserving substance that cannot be created or destroyed, there are no energetic crises or crises involving energetic resources when one takes the first law of thermodynamics into account. As per the second law of thermodynamics, the energy potential of an energetic processor is consumed to move this process, rather than decaying or being converted, as is commonly stated. The availability of energy, or "available energy," is what actually dictates any thermodynamic process, and this is measured by "exergy."

Energy as a market good is not valuable in studies of energetic optimization of industrial processes; rather, it is exergy, or the potential or accessible energy, that has value. Nonetheless, because they are more widely understood, the phrases energy and efficiency are still used in everyday speech. Therefore, it is imperative that this new evaluation process be organized and systematized, starting with the development of application methodology and the defining of new concepts.

As we have seen, energy efficiency is becoming increasingly important in the oil industry, especially in light of the energy revolution and the economy's significant decarbonization. In this case, reducing energy use and increasing energy efficiency are two of the most crucial actions that need to be done to accomplish sustainability goals and decrease the industry's negative environmental effects. Because of this, we aim to give a thorough introduction to thermodynamics, the theoretical foundation for analyzing energy and resource efficiency in industrial facilities—especially those involved in oil production.

Therefore, this book's main goal is to methodically arrange and clarify the ideas pertaining to the central topic of energetic efficiency while firmly establishing the progress of contemporary thermodynamics as the foundation for our investigation. Our goal is to close the knowledge gap that exists between engineering theory and practice by giving engineers and practitioners a complete toolkit for choosing, describing, and assessing plant installations and equipment from an integrated perspective. We want to give methods and analytical tools that elucidate the theoretical foundations and enable their practical implementation in industrial contexts, viewed through the prisms of thermoeconomics and energy systems engineering. In the end, this book aims to further the worldwide conversation on sustainable energy practices by promoting an approach to resource management and energy usage that is more ecologically friendly, accountable, and efficient.

To conclude, we give a set of methods for the analysis of industrial facilities with the aim of improving their efficiency, as instruments for putting the ideas and presumptions covered throughout the work to use. In order to do this, specific approaches that are recommended for application in projects based on these contemporary theories of thermoeconomics and energy systems engineering must be developed and illustrated.

The Fig. 1.1 represents an exaggerated illustration of an efficient offshore production facility concept and in summary, the future of energy optimization and efficiency is a multidisciplinary project that combines scientific, economic, and social concepts. It is not merely an engineering problem. Through furthering our comprehension and utilization of these ideas, we clear the path for a future in energy that is both equitable and sustainable. Furthermore, energy efficiency and other technologies like Carbon Capture, Utilization, and Storage (CCUS) are essential, particularly for the oil sector, to maintain the long-term appeal and sustainability of oil in the face of climate change and profound decarbonization.

1.1   Energy, Entropy, Heat, and Work

In thermodynamics, some of the most essential ideas include energy, entropy, heat, and work. In this chapter, we will explore these terms and their relationships in detail.

The property of a system to perform useful work is directly correlated to its level of energy. As we have seen, it can exist in a variety of manifestations, including kinetic energy, potential energy, and internal energy, among others. According to the

Fig. 1.1 An illustration of an efficient offshore oil production facility

first law of thermodynamics, the total amount of energy in a closed system is always the same. Energy cannot be generated or destroyed; rather, it can only be changed from one form into another.

A system's entropy can be thought of as a measure of how disordered or random it is. It is a property that increases with the passage of time and is connected to the quantity of energy that cannot be turned into work. According to the second law of thermodynamics, the total entropy of a closed system either continues to increase through time or stays the same, but it can never decrease.

Heat is a sort of energy carrier. It can be transferred from one object or system to another when there is a temperature difference between them. It is measured in energy units as Joules or calories, and it occurs when energy moves from a higher temperature to a lower one. This flow of energy is known as heat transfer. Heat can be converted into work, and vice versa, but not all heat can be converted into work due to this flux limitations and the ones imposed by the second law of thermodynamics.

Work is also an energy carrier. It is the transfer of energy from one system to another, which results in a change in the energy of the system that does the work. This change in energy is known as the work done. It is measured with energy units as Joules, as well. Work can be done by or on a system, and it can be positive or negative, depending on the direction of the energy transfer.

Through the application of the first and second laws of thermodynamics, one can formulate a mathematical representation of the connection that exists between these two words. According to the first law of thermodynamics, the difference between the heat that is added to a system and the work that is done by the system is equal to the change in the system's internal energy. The second law of thermodynamics indicates that the total entropy of a closed system either grows or stays the same, and it also states that the efficiency of a heat engine cannot be greater than the Carnot efficiency, which is dependent on the difference in temperature between the hot reservoirs and the cold reservoirs.

When it comes to the analysis and optimization of energy systems, having a firm grasp on the connections that exist between energy, entropy, heat, and work is absolutely necessary. The principles of thermodynamics provide a framework for designing efficient systems that maximize the use of available energy while minimizing waste.

1.1.1 Energy

The study of thermodynamics is predicated, in its most fundamental sense, on the concept of energy, which is distinguished by its propensity to produce effects and the potential from which all elemental interactions begin.

As mentioned before, on a macroscopic scale, it is possible to differentiate between two separate forms of energy: kinetic energy and potential energy. On the other hand, "thermal internal energy," which is inherent to a system, cannot be determined at this scale. When we put thermodynamics into practice, we frequently focus our attention on this particular type of energy, which is both a property of the system and a function of thermodynamic state.

Energy is the driving force behind everything in the universe, according to Adrian Bejan's interpretation. All the changes that are evident in our world are driven by its progression from one location to another. Kinetic energy, potential energy, and thermal internal energy make up the essential building blocks of thermodynamics. These three forms of energy provide a framework for comprehending how energy moves and is transformed.

Within this framework, considerable emphasis is placed on the principle of energy conservation. According to the first law of thermodynamics, energy cannot be generated nor destroyed; rather, it can only be transformed from one form to another. This is the only way energy can be changed. This concept has far-reaching ramifications, which affect everything from the design of devices that are efficient in terms of energy use to the study of universal evolution.

In summary, thermodynamics, also known as the study of the flow of energy and its transformations, centers on energy as its primary subject. This fundamental idea, which is supported by the principle of energy conservation, acts as a guide to direct our comprehension of all occurrences in the natural world. The insights provided by Adrian Bejan illuminate energy's critical role, enhancing our understanding of the physical laws governing our world and the universe.

1.1.2 Heat and Work

Heat and work are two forms of energy transfer that play an essential role in thermodynamics. In other words, they can be seen as energy carries. Work is the term used to describe the flow of energy that is caused by a potential difference that is not of a thermal nature, whereas heat is the term used to describe the flow of energy that is caused by a difference in thermal potential. Work is a form of energy transfer that takes place when a generalized force acts over a generalized displacement, while heat is a transfer of energy caused by a difference in temperature. Heat also transfers entropy, in addition to energy, and can be expressed analogously to work as the product of an intensive property (absolute temperature) and the variation of an extensive property (entropy).

As mentioned before, heat is the form in which energy flows due to a thermal potential difference, while work is the form in which energy flows due to a non-thermal potential difference. In other words, work is an interaction of energy transfer in which a generalized force acts along a generalized displacement, while heat is an interaction of energy transfer motivated by a temperature difference. We will see that heat also transfers entropy, in addition to energy, and that heat can be expressed similarly to work, as the product of an intensive property (absolute temperature) by the variation of an extensive property (entropy).

It is important to note that heat and work are not properties of a system, but are generally present, crossing its boundary, when the system undergoes a change of state. That is, they are not properties, but interactions that occur over time during the system's history. In summary, heat and work are interactions that occur at the system's boundary, through which there is an exchange of energy. Therefore, work and heat are forms of energy "in transit," and it is absurd to say that a body "possesses work" or "possesses heat." From a mathematical point of view, heat and work are line functions, whose differentials are non-exact (unlike the differentials of thermodynamic properties).

The following can be defined in more basic terms:

- Heat is the way in which energy flows due to a difference in thermal potential.
- Work is the way in which energy flows due to a potential difference that is not of a thermal nature.

The First Law of Thermodynamics:

As we have seen, according to the first law of thermodynamics, which is sometimes referred to as the law of energy conservation, the amount of energy that is contained within an isolated system remains the same. To put it another way, energy cannot be created or destroyed; rather, it can only change forms. The first law can be stated in a mathematical form as follows:

$$\Delta U = Q - W$$

where ΔU represents the change in the system's internal energy, Q represents the heat that the system has absorbed, and W represents the work that has been done on the system. This equation demonstrates that heat and work both transfer energy and that there is a corresponding change in the system's internal energy whenever there is a transfer of energy in the form of heat or work either into or out of the system.

The Second Law of Thermodynamics:

According to the second law of thermodynamics, the amount of total entropy always increases over the course of time within a closed system. The second law of thermodynamics can be stated in mathematical terms as follows:

$$\Delta S \geq Q/T$$

where ΔS denotes the change in entropy of the system, Q indicates the amount of heat that was absorbed by the system, and T is the temperature in its absolute form. This equation demonstrates that the only thing that can transport entropy from one system to another is heat, and that the direction of spontaneous change is always in the direction of an increase in entropy. To put it another way, heat flows from a hotter body to a cooler one and the overall entropy of the system as well as its surrounds always increases or, in the case of a special and reversible process, it remains the same.

Chapter 2
Exergy Concept

The exergy of a system in a defined state is the maximum amount of theoretically useful work that can be obtained when the system interacts with its environment in order to go from the defined state to mutual thermodynamic equilibrium, with the heat interaction performed exclusive heat with the environment. This led to the development of its etymology. The word "exergy" comes from the Greek language. It is formed by combining the prefix ex-, which means "external" or "outwards," with the segment ergon, which can represent either "force" or "work." In other words, the definition of exergy is "force out" or "force that can be extracted."

In the same way that we may understand that the exergy of a system in a given state is the smallest amount of theoretically useful work required to generate the amount of matter in the system from substances that are already present in the environment and get the matter to the stated state, we can also understand that the exergy of a system in a defined state is the maximum amount of theoretically useful work that is required.

Therefore, we can make the assertion that exergy is a property of the system that is coupled with its environment. This environment needs to have its parameters specified. The exergy of a system can be thought of as a quality that provides a quantitative measurement of the state of the system in relation to the condition of its surroundings. If a system is already in equilibrium with the environment, then the exergy will have reached its lowest possible value, which is zero.

However, if we are unable to implement a property that represents theoretical work, why are we interested in it? We must use the definition of another property to provide the solution to this question: it is entropy.

Entropy, abbreviated as "S," is a thermodynamic property that indicates the degree to which particles in a physical system are disorganized. This disorder arises, for instance, whenever there is a change in its temperature, which, as a consequence, also results in a change in the agitation of its molecules. The phase shift, represented in the Fig. 2.1, is an additional example that is valid. When we see ice melting all by itself, we know that the temperature at which it melts is going to be the same

M. V. da Silva Neves and A. F. Flutt, *Energy Efficiency in Oil Production*,
SpringerBriefs in Applied Sciences and Technology,
https://doi.org/10.1007/978-3-031-54274-9_2

every time. Therefore, all the heat that is supplied by the surrounding environment
is utilized to alter the bonds that are present between the water molecules that make
up ice. This results in a transition from the more structured and rigid arrangement
of ice to the more chaotic form of flowing water. When heated, water undergoes a
change that can be described as an increase in its entropy.

Consider another scenario, shown in Fig. 2.2, in which there is a fixed volume, and
the mass of the liquid is known, but the system is hermetic and adiabatic, meaning
there is no heat exchange with the surrounding environment. When we add heat to
water, or Q, we first cause its temperature to rise, and then we witness the water
transform into steam because of this transformation. If we keep adding heat to the
system, there is a good chance that the pressure and temperature of the system will
eventually rise. Water goes through several different physical changes as it transforms
from a subcooled liquid to a superheated vapor, all of which contribute to an increase
in the amount of disorder that is present in its molecules. To put it another way, we
saw a rise in the entropy of the system.

Now, if we were to do work by expanding this mass of superheated steam in a
turbine to the state in which the water was discovered initially, as represented in the
Fig. 2.3, we would note that it would not be feasible to convert all of the energy
that was supplied to the system by the transfer of heat into productive work. This is
because it would not be possible to return the water to the state in which it was found

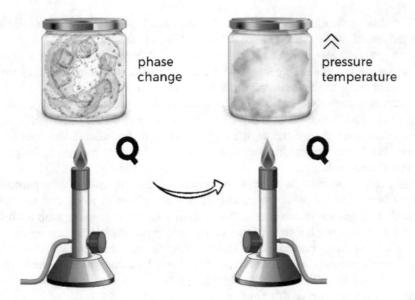

Fig. 2.2 Supply of heat to water in a constant volume adiabatic system

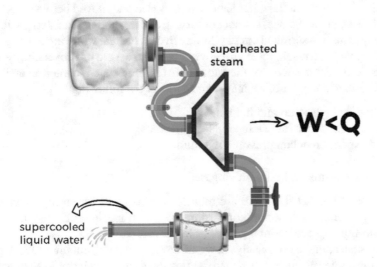

Fig. 2.3 Generation of work (W) from superheated steam

initially. This is due to the fact that some of that energy was "spent" on the process of entropy generation. To put it another way, the full capability of providing thermal energy to perform work that is beneficial has been eliminated. Because of this, we are able to say that the process cannot be reversed.

The mere realization of this truth is all that is required for us to comprehend the significance of the first statement of the second law of thermodynamics.

At a given temperature, it is not possible to extract thermal energy from a system and transfer all of this energy into mechanical work without making any changes to the system or the environment in which it is located (Kelvin's statement).

According to the second law of thermodynamics, even while it is feasible to convert all work into thermal energy, it is not possible to do the opposite. In turn, the measurement of an increase in entropy enables us to measure the fraction of energy that, in thermodynamic transitions at a certain temperature, can no longer be converted into work.

As a result, we are able to draw the conclusion that the exergy of a particular system in a certain state is the maximum amount of useful work that might be obtained through the interaction of that system with its surroundings up until the point where they reach a state of thermodynamic equilibrium with one another, if there were no entropy creation. If we measure the amount of entropy that is produced by a process, we can also determine the extent to which that process's capacity to perform productive work has been diminished. In other words, we can assess the degree of irreversibility of this process.

Because of this, there is a distinct difference between energy and exergy. It is true that energy is constantly preserved, and it is also possible for it to be stored and transmitted. Exergy, on the other hand, can be saved and moved between locations. Nevertheless, in most cases it is not conserved, and it is possible to destroy it.

The systems that surround our day-to-day lives in the real world are never going to be perfect, and as a result, there will always be some degree of irreversibility. There are a few different methods for the entropy that are connected to the irreversibilities of a process. Among these, we may name some of the following:

- friction (in both solids and fluids);
- heat transfer with a finite temperature difference;
- rapid expansion or compression of a fluid;
- combustion;
- spontaneous mixing of dissimilar gases.

In light of the fact that there are multiple mechanisms for entropy generation, it is only natural to draw the conclusion that these mechanisms can induce the irreversibility of processes to a greater or lesser degree.

For instance, in the prior analysis, we discovered that the amount of work generated by the turbine would be lower than the amount of energy that would be transferred to the system as a result of heat transfer. We are now able to identify a number of the contributors to this irreversibility, including the temperature difference between the thermal energy source and the water, changes in the state of the water, friction induced by the passage of steam inside the turbine, rapid expansion of steam, friction in the bearings of the turbine, etc.

Let us now ponder the possibility of another hypothesis. In the last illustration, we have transferred energy to the system through the heat carrier. What would happen if, instead of adding heat, we transferred the same quantity of energy (but this time, electrical) to an electric motor?

If we were to use the shaft of this motor instead of the turbine, would we be able to generate a greater or lesser amount of useful work?

In the instance of the electric motor, we would most certainly be able to make more efficient use of the energy that was put in. This is due to the fact that, although there are additional factors that generate entropy in this process (such as bearing friction, winding heating, and so on), the amount of entropy generated is significantly lower. If we examine two different ways of converting energy—namely:

(a) the conversion of thermal energy into mechanical energy;
(b) the conversion of electrical energy into mechanical energy,

We find that the first process, which converts thermal energy into mechanical energy, is more irreversible than the second process, which converts electrical energy into mechanical energy.

We are able to draw the conclusion that if we have 10 megajoules (MJ) of electrical energy and 10 MJ of thermal energy accessible, the 10 MJ of electrical energy presents a higher "quality" of energy than the 10 MJ of thermal energy does. This is the case despite the fact that both sources have the same energy quantity. The reason for this is that through the existing conversion mechanisms, we are able to acquire a greater beneficial impact in the second kind of energy conversion.

When we examine our processes through the lens of the exergy notion, this is one of the many significant benefits that we obtain. We are able to evaluate the "energy quality" as a result.

The first law of thermodynamics, which established the principle of the conservation of energy (as we have seen in the first chapter), does not differentiate between the many forms and qualities of energy and instead presents an analysis that is entirely quantitative. Once you have a firm grasp of the second law of thermodynamics, you will be in a position to recognize the distinctions that exist among the characteristics of various processes.

The process of exergy destruction is intimately linked to the generation of entropy. For this reason, some types of energy reserves, although with the same magnitudes, have different potential of useful use. Exergy provides us with an appropriate assessment since it enables us to examine both the quantitative and the qualitative aspects of the transformations that take place in processes. This is made possible by the combination of these two laws of thermodynamics.

Exergy allows us to evaluate all the streams that cross a system and their respective energy potentials on a single basis, whatever their respective qualities. Let's look at a common scenario to better grasp what we're talking about.

Let's say that at the end of a journey that took place across multiple countries and involved spending money in a variety of currencies, the traveler had the following sum of money in their bank account: 1000 yen, 1000 dollars, and 1000 euros, as summarized in the Fig. 2.4.

However, we are aware that it is not possible for us to simply add them up because the purchasing power of each currency is different and is dependent on the current

Fig. 2.4 Monetary balance
of the trip

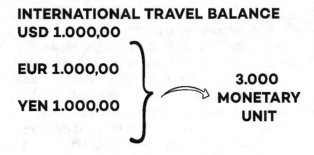

INTERNATIONAL TRAVEL BALANCE
USD 1.000,00

EUR 1.000,00

YEN 1.000,00

3.000 MONETARY UNIT

INTERNATIONAL TRAVEL BALANCE

USD 1.000,00 $\longrightarrow$ R$ 3.806,50

1 USD = R$ 3,8065

EUR 1.000,00 $\longrightarrow$ R$ 4.061,80

1 EUR = R$ 4,0618

1 YEN = R$ 0,0309

YEN 1.000,00 $\longrightarrow$ R$ 30,90

TOTAL:
R$7.899,20

Currency Exchange 2015/08/17

Fig. 2.5 Monetary balance of the trip on a single basis (BRL)

exchange rate. That is to say, each coin possesses its own "quality" in its own unique way. Therefore, even though in terms of quantity we have 3000 monetary units, it would not be right to add them together in this manner.

We need to tally them up on a single comparative basis to get an accurate understanding of just how much "purchasing power" we have left over from the remainder of the balance for that trip. In this particular circumstance, we change the entirety of the balance into Brazilian Reais (BRL). As a result, we would have had the monetary balance of Fig. 2.5.

After converting all of the amounts to Brazilian Reais, we were able to determine with pinpoint accuracy how much purchasing power we had left with the balance that was left over (BRL 7,899.20).

In a similar vein, we are able to reason in terms of the various types of energy and the attributes that each one possesses (Fig. 2.6). Assume for a moment that a manufacturing facility sends out the following types of energy products every second:

- 100 MJ of thermal energy in superheated steam at 14 bar and 285 °C;
- 100 MJ of thermal energy in water at 2.5 bar and 123 °C;
- 100 MJ of electrical energy,

with the environment in the following conditions: $P_0 = 1.013$ bar and $T_0 = 25$ °C.

Fig. 2.6 Energy production of an industrial process

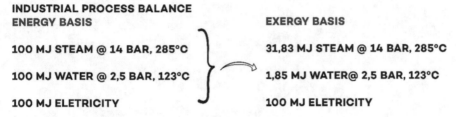

Fig. 2.7 Exergy production of an industrial process

Because every stream that crosses the boundary of this system possesses a unique potential to carry out useful work, we are aware that it is not sufficient to simply sum them all up. This is analogous to the situation with the monetary balance in the previous illustration. When compared to the thermal energy that may be obtained from steam and hot water, it is clear that electrical energy has the potential to be put to use in more noble commercial and home applications. And if we only consider this one facet of it, we see that it has the potential to be the most expensive.

In this sense, exergy is offered as a single basis for comparison, much to how the conversion of the balance to Brazilian Reais was presented as a single basis for comparison in the preceding example. Because of this, we are able to quantitatively increase each current while simultaneously taking into account the characteristics of its energy quality. This brings us to the results shown in Fig. 2.7.

As a result, the total amount of exergy in the system balance would be 133.68 megajoules, a quantity that is noticeably lower compared to the 300 MJ of energy. Because of this point of view, we are in a position to determine the extent to which we are capable of making use of our energy sources and to make the conscious decision to devote higher-quality and, as a result, more expensive resources to more worthwhile endeavors.

In conclusion, in studies on the energetic optimization of industrial processes, prospective or accessible exergy—rather than energy as a market good—is what has value. The phrases "energy" and "efficiency," on the other hand, are still employed in lay contexts because they are more widely understood. Therefore, it is crucial that this new way of evaluation be organized and systematic, starting with the creation of application methodology and the defining of new concepts.

Chapter 3
Energy Return on Investment and the Competitiveness of Oil

Energy Return on Investment (EROI) is a measure of the quality of various fuels, which calculates the ratio between the energy provided by a particular fuel to society and the energy invested in capturing and delivering that energy in its final or useful form, depending on the boundary of analysis. As Hall et al. (2014) point out, the EROI is expressed as an X:Y ratio, where investing one unit of energy Y results in X units of energy obtained. If the X:Y ratio is less than 1, there is no energy gain, meaning that there is an investment greater than the return obtained. So, the EROI plays a crucial role in assessing various energy systems. This principle calculates the ratio of energy derived from a specific source to the energy invested in its acquisition. The EROI serves as a guide to measure the energy efficiency of diverse sources and technologies and offers insights into their sustainability and economic viability (The Fig. 3.1 attempts to illustrate this abstract concept).

Numerous studies have analyzed the energy return for different sources (Gupta & Hall, 2011; Hall et al., 2014; Bhandari et al., 2015; Gupta, 2018; Wang et al., 2021). Two results are commonly found: (i) the EROI of fossil fuels, in general, is higher than that of renewable sources, and (ii) the EROI of fossil fuels has been decreasing over the years. However, although the EROI of fossil fuels is declining, most renewable energy alternatives have substantially lower EROI values. Thus, while there is a positive aspect of introducing renewable electricity as a high-quality energy vector, replacing fossil sources, there are some relevant challenges (Hall et al., 2014):

- Renewable electricity is less reliable and predictable than fossil fuels.
- Renewable sources are usually not energetically dense enough to displace fossil fuel investments through traditional market mechanisms.
- Electricity lacks transport, storage, and distribution infrastructure that is not yet sufficient to meet society's demands independently of fossil fuels.

Thus, from the perspective of EROI, the current energy transition faces the challenge of intentionally replacing higher EROI sources with lower EROI sources, which is precisely the opposite trend of previous energy revolutions (Smil, 2011).

M. V. da Silva Neves and A. F. Flutt, *Energy Efficiency in Oil Production*, SpringerBriefs in Applied Sciences and Technology, https://doi.org/10.1007/978-3-031-54274-9_3

Fig. 3.1 An illustration of the EROI concept

Petroleum currently leads the pack in EROI evaluations, with its high energy density and relative ease of extraction contributing to a significantly higher EROI than most renewable energy sources. As Hall et al. (2014) noted, petroleum's EROI ranges between 15:1 and 20:1, while solar and wind power typically fall between 5:1 and 10:1. This disparity means that for every unit of energy invested in oil production, we gain 15–20 units in return, as opposed to only 5–10 units from renewable sources.

The superior EROI of petroleum propels its competitive edge over most renewable energy sources. It is a highly concentrated energy source, cost-effective in extraction and transport, making it an attractive choice for various energy applications, from transportation to electricity generation.

Nonetheless, it's vital to recognize that the EROI of oil isn't static. As petroleum reserves deplete, extracting the remaining reserves becomes more challenging and energy-intensive, consequently decreasing its EROI over time and potentially undermining its competitiveness against renewable sources.

Moreover, we cannot overlook the environmental impact of oil extraction and combustion. Despite its EROI advantages, petroleum's negative externalities, including air and water pollution, greenhouse gas emissions, and ecosystem degradation, are substantial. Renewable energy sources, in contrast, have much lower environmental impacts, marking them as more sustainable and desirable options in the long run.

While the EROI is a valuable metric, it doesn't capture every aspect of the current energy transition, particularly the urgent need to reduce CO_2 emissions. Cleveland (1984) explains that the original application of EROI aimed at finding energy alternatives that could sustain economic growth while matching or exceeding oil's efficiency. Today's challenge lies in introducing energy alternatives that mitigate CO_2 emissions, despite their typically lower EROI (Hall et al., 2014). Thus, the selection of technologies that balance efficiency, economic return, and environmental impact mitigation must consider both EROI and CO_2 emissions.

Moreover, the traditional EROI evaluation approach, relying on calorific values to quantify direct and indirect energy input and output, has limitations. It only estimates the quantity of energy, disregarding energy quality, a crucial factor determining an energy source's societal utility. It also fails to provide a comprehensive view of the system's complexity, like labor, auxiliary services, and environmental inputs. To address these issues, physical approaches such as emergy and exergy analysis have been developed. These account for both the quality and complexity of the system, and other physical approaches include ExROI, Return of Exergy on Investment in Exergy, and minimal exergy return rates required by society.

In conclusion, although oil's high EROI grants it considerable competitiveness over most renewable sources, we must consider the long-term sustainability and environmental impacts of our energy choices. As we transition toward a sustainable energy future, we need to continue investing in renewable energy technologies and mitigating our current energy systems' negative impacts. The challenge of transitioning from high EROI sources to lower EROI sources demands careful planning and infrastructure investment to effectively integrate renewable sources into our energy system.

References

Bhandari, K. P., Collier, J. M., Ellingson, R. J., & Apul, D. S. (2015). Energy payback time (EPBT) and energy return on energy invested (EROI) of solar photovoltaic systems: A systematic review and meta-analysis. *Renewable and Sustainable Energy Reviews, 47*, 133–141. https://doi.org/10.1016/j.rser.2015.02.057

Cleveland, C., Costanza, R., Hall, C., & Kaufmann, R. (1984). Energy and the US Economy: A biophysical perspective. *Science, 225*(4665), 890–897.

Gupta, A. (2018). Energy return on energy invested (EROI) and energy payback time (EPBT) for PVs. In *A comprehensive guide to solar energy systems* (pp. 407–425). Elsevier. https://doi.org/10.1016/B978-0-12-811479-7.00021-X

Gupta, A. K., & Hall, C. A. S. (2011). A review of the past and current state of EROI data. *Sustainability, 3*(10), 1796–1809. https://doi.org/10.3390/su3101796

Hall, C. A. S., Lambert, J. G., & Balogh, S. B. (2014). EROI of different fuels and the implications for society. *Energy Policy, 64*, 141–152. https://doi.org/10.1016/j.enpol.2013.05.049

Wang, C., Zhang, L., Chang, Y., & Pang, M. (2021). Energy return on investment (EROI) of biomass conversion systems in China: Meta-analysis focused on system boundary unification. *Renewable and Sustainable Energy Reviews, 137*, 110652. https://doi.org/10.1016/j.rser.2020.110652

Smil, V. (2011). Science, energy, ethics, and civilization. In *Visions of discovery: new light on physics, cosmology, and consciousness* (pp. 709–729). Cambridge University Press. https://books.google.com.br/books?hl=pt-BR&lr=&id=BhcpiZN2MOIC&oi=fnd&pg=PR7&dq=SMIL,+Vaclav.+2010.+Science,+energy,+ethics,+and+civilization.+Cambridge:+Visions+of+Discovery:+New+Light+on+Physics,+Cosmology,+and+Consciousness,+R.Y.+Chiao+et+al.+eds.,+pp.+709-729+-+Cambridge+University+Press.&ots=GXdAwNtzzh&sig=3VJVb1XUG_dzgtaQ5Q9dZZwOoXc#v=onepage&q&f=false

Chapter 4
Energy or Exergy Analysis in Oil Production

Abstract A petroleum and gas production unit consists of a wide variety of components and deals with different forms of energy, such as chemical energy from fuels, electrical energy, and thermal energy for heating or cooling.

A petroleum and gas production unit consists of a wide variety of components and deals with different forms of energy, such as chemical energy from fuels, electrical energy, and thermal energy for heating or cooling. Each component serves distinct objectives and has specific production methods.

Conventionally, the determination of the energy efficiency of the unit is made in global terms or for each component. And, in general, the analysis is based only on the first law of thermodynamics, requiring the quantification of the various forms of energy transferred to and from the outside, but without considering the conditions. In overall terms, the calculation of efficiency consists of establishing a ratio between the amount of energy associated with the desired effect with a certain process and the amount of energy actually required to carry it out. The difference between these quantities of energy is considered loss.

On the other hand, the analysis on which the determination of the so-called exergy efficiency is based considers the thermodynamic conditions of the external environment to the unit in which the global process is performed or to the component in which a specific process is carried out. The calculation of efficiency then consists of establishing the ratio between the maximum reversible potential energy available or remaining in the process or useful effect obtained and the maximum possible capacity of transformation of the amount of energy required to achieve these objectives. The capacity of transformation of a certain amount of energy, regardless of its form, is measured relative to the external environment and, as we have seen previously is called exergy. Exergy establishes, in other terms, a common basis for comparing the transformation capacity associated with different forms of energy.

The results obtained because of exergy analysis have additional significance in relation to those provided by energy analysis, as they offer more comprehensive conditions for decisions concerning changes in project or operational conditions of the units.

© The Author(s), under exclusive license to Springer Nature Switzerland AG 2024   23
M. V. da Silva Neves and A. F. Flutt, *Energy Efficiency in Oil Production*,
SpringerBriefs in Applied Sciences and Technology,
https://doi.org/10.1007/978-3-031-54274-9_4

As Szargut pointed out, the application of exergy analysis in the oil production process can lead to more efficient use of energy and fewer environmental impacts. This is because exergy analysis considers the quality of energy and provides more detailed information about the losses and the potential for improvement in energy conversion processes.

Adrian Bejan's constructal theory can also be applied to oil production systems, where the aim is to design a system that minimizes energy consumption and maximizes the flow of oil and gas. This theory states that the natural tendency of any system is to evolve toward the most efficient configuration possible. In oil production, this means that the design should allow for the most efficient flow of fluids and the use of energy, while minimizing losses and environmental impacts.

The importance of exergy analysis in the energy industry, as highlighted by Antonio Valero, emphasizes its ability to facilitate a thorough evaluation of energy quality and the potential improvements in conversion methods. This viewpoint holds particular significance within the oil production sector, given the industry's substantial energy requirements and the imperative to address environmental considerations through enhanced energy resource management practices that are both efficient and sustainable.

Bringing the exergy analysis to the oil production processes, in a seminal investigation carried out in 1997, Ghorzi's master's thesis at UNICAMP presented a comprehensive exergetic analysis of primary oil separation plants. The study conducted a comprehensive analysis focusing on distinct operations inside primary oil separation units, with a particular emphasis on evaluating the exergetic efficiency of individual processes.

Ghorzi's research also encompassed a comprehensive literature assessment, incorporating terminology and perspectives from prominent exergy scholars. The discussed subjects encompassed essential themes such as exergetic analysis, exergetic efficiency, and the notion of a reference environment. The review established the foundation for the use of these techniques in primary oil separation plants.

The examination of specific case studies was conducted utilizing a computational simulator, with a particular emphasis on physicochemical phenomena. The implications derived from these investigations emphasize the significance of the findings pertaining to primary oil separation facilities. The author suggests for the incorporation of exergetic analysis as a major design criterion in the building of future primary separation facilities. He emphasizes the potential of exergetic analysis to improve both efficiency and sustainability.

4.1   Exergy Analysis

In the complex and dynamic domain of energy efficiency, namely within the realm of petroleum production, the notion of exergy analysis emerges as a pivotal instrument for enhancing energy consumption and mitigating inefficiencies, as we have seen above. This particular segment of the book explores the core principles of exergy analysis, a methodology that surpasses conventional energy assessment by not only quantifying energy but also assessing its quality and capacity to perform useful work. The application of exergy analysis, which is based on the principles of the second law of thermodynamics, offers a more extensive comprehension of energy conversions and dissipation. Consequently, it proves to be an indispensable methodology for optimizing the efficiency oil production facilities.

The following subsections will elucidate the principles and methodologies of exergy analysis, with a focus on its practical implications in the petroleum industry. Starting with the crucial concept of the reference environment, we will explore how exergy is influenced by external conditions and proceed to examine the methodologies for exergy computation. This introduction sets the stage for a thorough exploration of exergy analysis, an essential tool for achieving higher efficiency in the oil production.

4.1.1 Reference Environment

It is vital to evaluate the significance of creating a reference environment for exergy computation since a system's exergy potential is dependent on its external environment and because the environment's thermodynamic properties are typically not constant over time and space.

The establishment of the reference environment is a topic that can involve extensive discussions. They will, however, be simplified here because chemical exergy will essentially not be analyzed and physical exergy will play a significant role in the analysis.

There are essentially two different perspectives on how to identify the reference environment. One of them maintains that the environment must be selected for each situation evaluated, considering the temperature and environmental pressure in the system's boundaries, as well as the chemical potential of the substances present in the vicinity of the system in question.

This methodology ought to be applied in situations where the system under study exhibits notable operational fluctuations in response to environmental shifts (such as thermal and refrigeration machines) or in settings where the characteristics of the environment are markedly different from global averages, like polar or desert regions.

The other current maintains that the environment has a normal and fixed condition of temperature, pressure, and chemical composition, even if this condition does not coincide with the real one in every local or temporal situation, according to Ahrendts (1980), Sussman (1980), and Szargut et al. (1988e).

The temperature $T0 = 298.15$ K ($25\,^{\circ}$C), $P0 = 101.325$ kPa, and common chemical species in the Earth's crust, oceans, and atmosphere are taken into consideration in the Szargut et al. (1988) proposal for determining exergy.

In this book will make use of the second proposal. The case relates to oil production makes no claims to be statistically rigorous, despite the fact that some of the systems included in the model—such as thermal machines—are highly sensitive to changes in their environment. An evaluation that is both qualitative and comparative is the goal. Thus, for oil and gas production units situated at sea with average temperatures around the proposed value, an ambient temperature of 25 °C and an environmental pressure of 101.325 kPa, referring to sea level, are highly coherent with the scenarios to be evaluated.

4.1.2 Analysis Method

The first law of thermodynamics, which postulates the principle of energy conservation, is the most widely used method of thermodynamic analysis of a system. However, as Sect. 4.1 points out, this approach is purely quantitative and does not consider the various attributes of the energy conversion methods.

Consequently, it is necessary to assess the second law of thermodynamics in order to determine the characteristics of processes using the ideas of entropy and its generation provided by this postulate.

If the second law evaluates the generation of entropy, it also identifies the destruction of exergy and, therefore, an efficiency analysis based on the second law allows determining the capacity of the process effluents to perform work, enabling comparison with the capacity of the streams.

In other words, in every process, the final exergy is comparable to the initial exergy. This is the fundamental idea of energy analysis, which is covered in more detail in the section that follows.

The efficiency found for each of the two laws is utilized to combine them using the idea of real efficiency, according to Stephan and Mayinger (1998c). The highest efficiency attained by the maximum useful work (L_{ex}) converted from any energy source (Q) by an ideal reversible process is quantified by the efficiency of the first law.

$$\eta_I = \frac{L_{ex}}{Q}. \tag{4.1}$$

The efficiency of the second law measures the deviation between the actual work obtained (L) by the evaluated process and the maximum that would be obtained if the same process were ideal and reversible (L_{ex}).

$$\eta_{II} = \frac{L}{L_{ex}}. \tag{4.2}$$

The real efficiency is the combination of these two efficiencies. It simultaneously evaluates the quantity and quality of the useful energy obtained, through the quantification of the maximum work developed in an ideal process discounted from the real losses due to the irreversibilities of the process and the consequent generation of entropy.

$$\eta_{Real} = \eta_I \cdot \eta_{II}, \tag{4.3}$$

Thus, by solving Eq. (4.3) through Eqs. (4.1) and (4.2), we arrive at the expression for real efficiency:

$$\eta_{Real} = \frac{L}{Q}. \tag{4.4}$$

The Eq. (4.4) cannot yet be considered an efficiency on an exergy basis, as it evaluates the exergy available in the product, the real work (L), but does not identify the exergy of the fuel, the one that accompanies the heat (Q).

Therefore, to obtain exergy efficiency, it would be necessary to use the fuel's exergy, which in this case is that which accompanies the heat Q and is defined according to the equation:

$$E_Q = \left(1 - \frac{T_u}{T}\right) \cdot Q, \tag{4.5}$$

Therefore, the first definition for exergy efficiency is given by the relationship between the exergy of the product (E_P, in this case, L) and that of the fuel (E_F, in the evaluated situation, E_Q), according to Eq. (4.6):

$$\varepsilon = \frac{E_P}{E_F} = \frac{L}{E_Q} = \frac{L}{1 - \frac{T_u}{T} \cdot Q}. \tag{4.6}$$

As an illustrative example, the efficiency of an electric shower can be examined through the following comparison. According to the first law of thermodynamics, the electric shower demonstrates a high level of efficiency, nearing 100%. This is due to the fact that a significant portion of the electrical energy is converted into thermal energy, as evidenced by the noticeable rise in water temperature. However, it is important to note that the hot water produced by the electric shower is unable to perform the same amount of work as the initial electrical energy that was used to heat it. Consequently, the process exhibits a considerable degree of irreversibility. The principle of energy conservation was upheld; nevertheless, the quality of the energy experienced a decline, resulting in a fall in exergy. The second law of thermodynamics addresses the impossibility of generating an equivalent amount of electrical power through the utilization of hot water that was initially used to produce the heat. In practical applications, the exergy efficiency of an electric shower is nearly negligible (The Fig. 4.1 represents an illustration of this example).

In a facility composed of several systems and processes, the exergy method allows evaluating not only the total yield of the plant, but also the contribution of each portion to the degradation, throughout the entire process, of the initial exergy.

Fig. 4.1 An illustration of energy efficiency and exergy concepts

Hence, it is feasible to ascertain the process inside a facility that has the most negative impact on its efficiency, or even discern the extent to which variations in process performance contribute to the overall efficiency of the plant.

While, in this book we will not specifically address the economic aspect, it is worth noting that the utilization of the exergy analysis approach enables a sophisticated and efficient integration with economic evaluation, leading to the emergence of exergoeconomics.

Reference

Szargut, J., Morris, D. R., & Steward, F. R. (1988). *Exergy analysis of thermal, chemical, and metallurgical processes*. Springer.

Chapter 5
Energy Efficiency in Oil Production Facilities

Before diving into oil production facilities and identifying ways to make them more energy efficient, the reader may ask himself the following questions:

- "What is energy efficiency in an oil production facility?"
- "What should be optimized in your production processes?"
- "How should we measure its performance?"

Such questions are apparently simple; however, certainly, experienced professionals will have different answers. To start the chapter and have a more homogeneous view among readers, the first subsection of this chapter provides the concept of energy efficiency in oil production facilities to be used in this book. Next, we present the reader with basic information about oil production facilities and their production processes, to gradually provide information related to the main theme of the book, energy efficiency. Strangely, it is common to see professionals relating the increase in energy efficiency to the increase in the initial cost of investing in oil production facilities, which is not an inexorable reality. Thus, with this chapter, we aim to sensitize the reader to simple (energy efficiency) actions that have little impact on initial investment costs, or even reduce these costs. Naturally, such actions promote a significant reduction in operating costs, which are present throughout the life cycle of facilities.

5.1 Concept of Energy Efficiency in Oil Production Facilities

For oil production facilities, energy efficiency is the ratio between the energy contained in products and the energy expended to produce them (a definition that is close to the one that we have seen in the previous chapter related to EROI). Thus, optimizing energy efficiency in an industrial oil production and processing facility

© The Author(s), under exclusive license to Springer Nature Switzerland AG 2024

M. V. da Silva Neves and A. F. Flutt, *Energy Efficiency in Oil Production*,
SpringerBriefs in Applied Sciences and Technology,
https://doi.org/10.1007/978-3-031-54274-9_5

means producing the maximum number of products with the least possible energy consumption, assessing the economic, environmental and social impacts and benefits. This concept serves as the basis for the other subsections of this chapter.

5.2   Oil Production Facilities

In order for the reader to become familiar with the peculiarities related to oil production facilities, this section briefly presents some characteristics of offshore oil production facilities, known as Stationary Production Units (UEP) and their main systems. Although this chapter deals with any type of oil production and processing facility, whether offshore or onshore, as the production systems and processes are similar for both facilities, in this chapter we will only exemplify offshore facilities and their respective systems. There are several concepts of UEP. In Brazil, we basically operate three types: Fixed Units, Semi-submersible Units and Floating Processing, Storage and Offloading Units, as per the subsections below.

5.2.1   Fixed Units

Fixed platforms are stationary units limited to a water depth of approximately 150 m. They house production plants composed of specific modules (oil processing, water treatment, gas compression, power generation, accommodation and offices, etc.) installed on metal structures fixed to the sea floor. We can see an example of this type of installation in Fig. 5.1.

5.2.2   Semi-submersible Units (SS)

Semi-submersible Platforms are Stationary Units (represented in the Fig. 5.2) whose production plants are supported by columns connected to submarines (Pontoons) responsible for the ballast and buoyancy of the unit. There are platforms of this type operating in a water depth of more than 1500 m. Usually, these units are old converted drilling rigs.

Fig. 5.1 Example of Fixed Platforms operated in Brazil. Author: Tais Peyneau, Petrobras Image Bank; Pargo 1A and 1B twin platforms operating in the Northeast Pole in the Campos Basin

5.2.3 Floating Processing, Storage, and Offloading Unit (FPSO) Type Units

The Floating Processing, Storage, and Offloading (FPSO) concept of stationary units basically consists of an oil cargo ship converted into a production unit. This modality is an economical alternative when there are hulls available on the market, as we can see in the Fig. 5.3. These units have an anchoring system to the seabed that allows a certain degree of alignment to the direction of the wind and ocean currents according to the weather conditions. They also have the advantage of storing a significant amount of oil in their large tanks, which are also responsible for ballast and stability.

As the tendency of oil production in Brazil is to establish itself in increasingly deeper waters, most of the new projects are for Stationary Floating Units such as the SS and FPSO.

Fig. 5.2 Example of Semi-Submersible Platform operated in Brazil Author: Geraldo Falcão, Petrobras Image Bank; Semi-submersible platform P-19 in the Marlim field in the Campos basin

Fig. 5.3 Example of FPSO operated by PETROBRAS. Author: Geraldo Falcão, Petrobras Image Bank; Production vessel (FPSO) P-37 operating in the Marlim field in the Campos basin

Typically, the main systems related to energy are the following: (1) Oil Treatment and Transfer System, (2) Gas Compression System, and (3) Electricity and Hot Water Generation System.

5.2.4 Oil Treatment and Transfer System

The Oil Treatment and Transfer system receives the production flow from the subsea wells and processes them in order to specify the oil for the required terms of the relative amount of water and sediment (BSW—Basic Sediment and Water). Basically, this system separates the gas, oil, and water flows from the crude oil that arrives at the platform, forwarding each one to its destination.

The gas goes to the compression system, from where it can be exported for consumption on land or re-injected into underground reservoirs through subsea injection wells, or used for Gas Lift. Reinjection aims to postpone its use. The Gas Lift is one of the most common solutions for lifting oil from the producing wells to the treatment system. In addition to these destinations, normally part of the gas is treated to be used as fuel for thermal machines and furnaces in the installation itself.

Oil is sent to transfer pumps, being exported to other units, to shuttle tankers or to land.

The produced water goes to a specific treatment system, after which the residual oil is sent back to the oil treatment and the clean water is discarded. This water can also be sent to an injection system through subsea injection wells and used to maintain the pressure of the geological formation. This procedure is of great importance for the conservation of the productive potential of the field over the years of operation.

The oil treatment system, represented in the Fig. 5.4, separates the gas, oil and water streams initially through a thermo-physical decantation process. The feasibility of using this process stems from the great difference between the mass densities of water, oil and gas. The fractions of water and gas that are solubilized in the flow are more difficult to separate. For greater efficiency of your separation process, the streams are heated. The thermal energy responsible for this heating normally comes from the electric power generation system, through water heated in the Waste Heat Recovery Units (WHRU). These units recover heat from exhaust gases from turbogenerators or furnaces, which also use the gas produced as fuel.

A part of the water still remains emulsified in the oil load, and it is not possible to separate it simply by this thermophysical process. An electrostatic treater is then used to make the emulsified droplets come into cohesion and in this way, the water decants and is separated by the bottom of the vessel.

A small amount of residual gas is separated in the atmospheric separator vessel, which reduces the load arrival pressure to a value close to that of the atmosphere, causing the fraction of gas still present to separate through the upper part of the vessel.

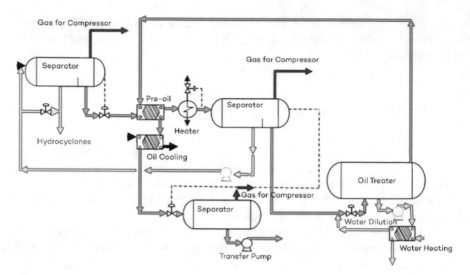

Fig. 5.4 Typical oil treatment scheme

5.2.5 Gas Compression System

The Gas Compression System, illustrated in the Fig. 5.5, comprises discharge compressors and vapor recovery units (VRU). These units receive the gas separated from the cargo in the oil treatment system.

The largest amount of gas goes to the high compression system. This system is composed of a battery of centrifugal compressors in parallel. Typically, each compressor consists of three stages configured in series. These machines are additional sources of thermal energy, generally not used, due to the high temperatures reached by the compression effect. Intermediate coolers are coupled to them to reduce the temperature between each process stage. The condensate formed as a result of compression and cooling is removed in vessels and redirected to the beginning of the treatment process.

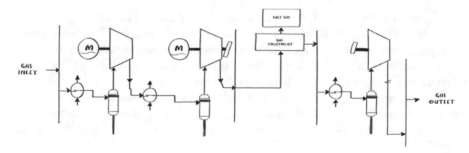

Fig. 5.5 Diagram of the three process stages of the main compression system

In the configuration of the example used here, these centrifugal compressors are driven by electric motors. These motors are among the unit's biggest consumers of electricity (from 30 to 50% of total consumption). For flow control and to avoid recycle or valve restrictions, in some cases frequency variators are used to adjust the rotation of the electric motor to the required flow; in other cases, hydrodynamic variators are used, which make the reduction in a similar way to fixed mechanical reducers, but with the speed adjusted in a hydrodynamic way. When correctly applied, these devices also avoid wasting energy. Flow control by flow recycle or valve throttling (usually suction) are energetically unfavorable.

After compression, the gas can proceed to:

- Export, typically to natural gas processing units on land, which process the gas according to pre-established specifications and send it for consumption.
- Gas Lift, which is one of the most used lifting techniques in oil production. The gas is injected into the production column through its annulus, reducing the weight of the oil column, therefore increasing the pressure differential between the oil reservoir and the producing well, thus facilitating the lifting and flow of the load.
- Fuel Gas. The gas is treated, basically separating the condensate, to be used as fuel gas in the turbines of the electric power generation system.
- Reinjection. The gas is reinjected into the geological formation in injection wells, postponing its production in case of lack of demand and/or infrastructure for handling the product.
- Other purposes. There are other purposes for gas, known as Non-Energy Use of Gas, such as: deoxygenation of injection water, motive gas, etc. But the demands of these systems are smaller and do not always exist in the projects.

The low pressure gas, mainly coming from the atmospheric separator, has a very low pressure, insufficient to send it to the high compression system. It is therefore compressed through the Vapor Recovery Unit (VRU) to the suction pressure of the high compressors. The VRU is made up of a screw compressor (Positive Displacement) driven by an electric motor and works as a High Compression Booster.

5.2.6 Electricity and Hot Water Generation System

The Electricity and Hot Water Generation System basically corresponds to the facility's turbogenerators. Commonly, this system is formed by aeroderivative turbines, known as Gas Generators. Turbines supply enthalpy in the form of flue gases and air at high temperature and pressure to industrial power turbines (see Figs. 5.6 and 5.7). The latter are coupled directly, or through speed reduction boxes, to the electric power generators.

The Fig. 5.8 shows the typical arrangement of a turbogenerator, and Fig. 5.9 illustrates the half of the electric power generation module in an offshore unit. Part of the thermal energy of the exhaust gases is used in the WHRU. These units are basically heat exchangers. The hot water circuit receives energy to heat the water

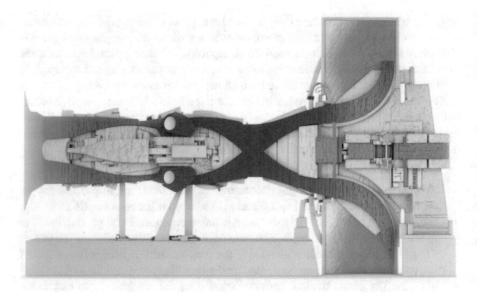

Fig. 5.6 Turbogenerator

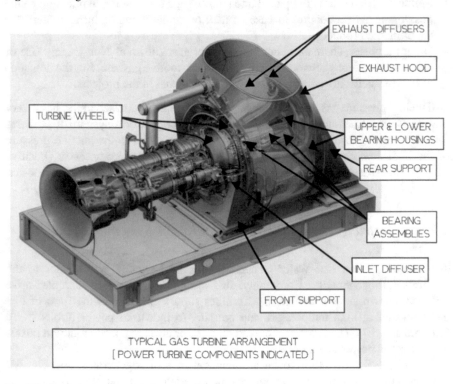

Fig. 5.7 Typical arrangement of a gas turbine

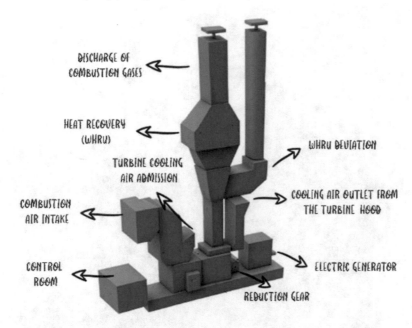

Fig. 5.8 Typical arrangement of a turbogenerator with water heating through WHRU

from these stoves. As described in Sect. 5.2.4 on the Oil Treatment and Transfer System, hot water is mainly used to heat the incoming load from the producing wells, in order to facilitate the separation of the 3 fluid streams (gas, oil, and water). In some cases, there may be additional heating of water in furnaces. Such heating consumes the gas itself produced.

5.3 Flow Machines (Pumps, Compressors, and Turbines)

In the context of energy efficiency, the main attention lies on pumps, compressors, and turbines, which are classified as primary flow machines. These machines have the capability to either use or generate energy. Therefore, it is imperative to incorporate crucial data in the evaluations of energy efficiency, such as the calculation techniques for power generation or consumption, as well as the customary thermodynamic and mechanical efficiencies associated with these devices. Additionally, it is worth mentioning that these machines release heat into the surrounding environment. Hence, it is imperative for any analysis to evaluate the capacity to utilize the aforementioned dissipated thermal energy.

Fig. 5.9 Half of the electric power generation module (2 Turbogenerators)

5.3.1   Conceptualization and Classification

Flow machines refer to a category of equipment through which a fluid continuously passes. These devices exhibit energetic interactions with the fluid that traverses them, facilitating the passage of energy between the machine and the flow, depending on the specific characteristics of the device.

There are two primary classifications of flow machines that have been identified: The first category of machines, known as operator machines, are responsible for transferring power to a fluid flow. Conversely, driving machines belong to the second category and are designed to obtain power from a fluid flow.

5.3.2 Machine–System Relationship

The choice of analytic approach depends on the characterization of the flow type involved, taking into account the unique behavior of the fluid: compressible (gases) or incompressible (liquids). The equipment utilized in each situation exhibits distinct characteristics that are specifically designed to fulfill their respective tasks. An endeavor is undertaken to offer a cohesive approach in addressing collections of machines that possess comparable geometric qualities and perform similar functions.

Irrespective of the specific type of flow machine being examined, it is important to acknowledge a fundamental principle regarding its behavior, which is often disregarded. This principle, as formulated by Rodrigues (1991), states that the performance of a flow machine is influenced not only by its design and adjustment characteristics, but also by certain parameters associated with the service it performs. These parameters include inlet pressure, inlet temperature, outlet pressure, and the nature of the fluid being processed. The aforementioned parameters function as the input data for any analysis of energy efficiency of flow machines.

This suggests that the analysis of a flow machine necessitates consideration of its system as a whole, rather than examining it in isolation. The interaction between many factors determines the resulting operating characteristics, such as flow rate, power requirements or generation, and stress levels.

5.3.3 Thermodynamic Analysis

In order to achieve practical objectives, a simplified thermodynamic analysis that evaluates global flow parameters is adequate for purposes such as performance testing and specification/selection.

The formulation of the 1st Law of Thermodynamics applied to a flow machine incorporates the introduction of the following assumptions:

There are four key aspects to consider in this context. Firstly, the concept of representation by control volume with an entrance and exit is important. Secondly, the assumption of uniform characteristics and invariance over time along a flow section is significant. Thirdly, the notion of steady-state operation is crucial. Lastly, the condition of constant mass flow rates being same at the entry and exit of the machine is noteworthy.

Hence, the Conservation of Energy Principle for a flow machine operating in a condition of equilibrium can be expressed as follows: "The sum of energy flows relative to the machine is zero."

The concept of specific energy in fluid dynamics is characterized by its definition as:

$$e = u + Pv + \frac{c^2}{2} + gZ \tag{5.1}$$

The initial component on the right side of the equation indicates the internal energy of the fluid inside a specific segment of the flow, which is associated with the molecular arrangement of the substance. The other terms correspond to the primary motion of the fluid mass, specifically referring to the mechanical energy of the fluid.

Operator Machine: performing work on the flow:

$$w - q_r = e_1 - e_2 \tag{5.2}$$

Driving Machine: obtaining work from the flow:

$$w + q_r = e_1 - e_2 \tag{5.3}$$

5.3.4 Thermodynamic Efficiency

The deterioration of energy in a flow is a result of the irreversible transformation of mechanical energy into internal energy caused by dissipative phenomena such as fluid friction and turbulence.

Dissipative effects manifest themselves within flow machines. In the case of an operator machine, the energy supplied by the driver is predominantly transferred to the fluid in the form of mechanical energy. However, a portion of this energy is converted into internal energy, leading to an increased energy requirement for a given task. The phenomenon observed in a driving machine is analogous, except in the opposite direction, where the decline in mechanical energy of the flow leads to a decrease in the amount of deliverable energy within specific operational circumstances.

The quantification of the non-ideality within a machine is achieved through the concept of thermodynamic efficiency. A theoretical machine model, which is not subject to energy deterioration, and operates under identical service circumstances as the actual machine, functions as a benchmark. Efficiency is thereafter calculated as the quotient of work per unit mass seen in both instances.

In the context of an operator machine:

$$\eta_{\text{th}} = \frac{w_{\text{ideal}}}{w_{\text{real}}} \tag{5.4}$$

For a driving machine:

$$\eta_{\text{th}} = \frac{w_{\text{real}}}{w_{\text{ideal}}} \tag{5.5}$$

In the context of an operator machine, energy degradation refers to the increased work demand imposed on the system's characteristics. Conversely, in a driving machine, energy degradation leads to a reduction in the amount of work delivered by the system's predetermined set of characteristics.

At this stage, it is necessary to make some implied comments regarding the definition of thermodynamic efficiency. The determination of ideal work is contingent upon the establishment of an idealized process, wherein energy degradation is absent, and which is executed under identical service characteristics as those observed in the corresponding real process. While it is possible to determine the ideal work using analytical means, the measurement of real work can only be achieved through experimental methods. On the other hand, possessing knowledge of the anticipated thermodynamic efficiency for a machine operating under particular service conditions enables the estimation of the corresponding actual work. This estimation is derived from the analytical calculation of the ideal work and the subsequent application of this efficiency. Performing this task is a common practice among users of flow machines.

5.4   Application of Exergy Analysis to Oil and Gas Production Facilities (External Observer Proposal)

An oil and gas production facility can be described as an open system whose input is a mixture of crude oil, gas, water, salt, sand, among others, and the output (products) is represented by streams of natural gas, petroleum specified, and water. Its driving force can be obtained in the following ways:

- Burning part of the gas produced;
- Burning of other types of fuels external to production, typically diesel oil;
- Electricity from a concessionaire or generated in another installation.

In the focus study of this book, the gas produced will be the only fuel source evaluated, as it is actually the main energy input of large marine production facilities.

The mixture that arrives from the wells is essentially two-phase (liquid + gas), so its transport to the continent cannot be done with common equipment. It is necessary to separate the gas from the liquid and pressurize both—the gas by compressors and the liquid by pumps, to transport them through specific pipes to their destination. Currently there are multiphase pumps for moving two-phase flow; however, this equipment still has significant flow and pressure limitations and is still in the technological development phase and, therefore, will not be considered in this study.

There is also a portion of the gas produced that can be used for gas lift as a method of increasing production and/or for reinjection in order to postpone its production in situations where there are no resources for movement or demand for this gas.

As already explained, the exergy analysis of a system takes into account the exergy of the effluents, the exergy of the input streams, the driving exergy of the process, irreversibilities, and losses to the environment. The correct use of the values found, as terms of the equations involved, requires precise definitions.

For oil and gas production facilities, the objective is to maximize production, with the minimum possible delivery of exergy to the system, that is, with the lowest generation of irreversibilities and lowest losses to the environment, with the aim of ensuring that Most of the exergy delivered to the system, in the form of fuel, goes with its products.

The exergy efficiency of an oil and gas production facility must be a number with the following characteristics:

- must be between zero and one (or zero and 100%), so that it is possible to compare its value with that of other installations;
- it must grow with the decrease in entropy generation (that is, with the decrease in the irreversibilities term) and with the decrease in losses to the environment, so that when these portions are reduced, efficiency increases.

For an observer external to an oil and gas production facility, who only sees the inlet and outlet currents, this facility can be represented, in terms of exergy flows, as shown below (Fig. 5.10).

As in an oil and gas production facility, the chemical reactions to separate the incoming stream (crude oil) from the outgoing stream (separated oil and gas + water and sediment) are minimal, the chemical exergy of these components does not need to be accounted for, as chemical transformations are negligible. Therefore, the model to be developed in this work will only take into account the physical exergy of the oil inflow and outflow streams. Although physical exergy also encompasses potential exergies and the kinetics of currents, these will be ignored in the analyses to be established in this dissertation, considering for physical exergy only the portion that the scientific literature considers as thermomechanical exergy. Therefore, whenever there is reference in this study to physical exergy, strictly speaking only thermomechanical exergy is being evaluated.

However, there is a term that, although the external observer cannot see it (it does not cross the border of the installation's control volume), is essential for the analysis. The exergy consumed by auxiliary processes (WAux), such as: injection of chemical products; sea water capture; exergy consumption of accommodation, offices and workshops; between others. The exergy of auxiliary systems is basically that consumed in the aforementioned systems, which do not transfer exergy to the products. By identifying this consumption, it will be possible to evaluate its contribution or weight in the use of the unit's fuel exergy.

Thus, the graphic representation of exergy flows in the control volume of an oil and gas production facility is identified in Fig. 5.11.

Therefore, to calculate exergy efficiency it is necessary to identify the exergy of the fuel and that of the products.

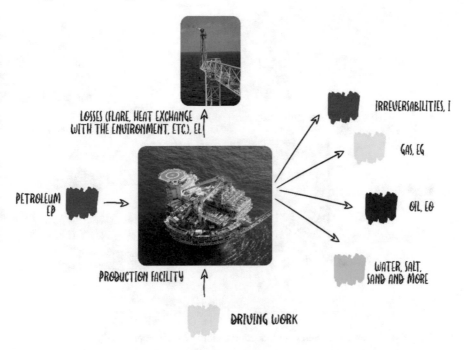

Fig. 5.10 Exergy flows in oil and gas production facilities. Where E_{pet}, physical exergy of the crude oil arriving at the installation; E_L, exergy lost to the environment, through flare burning, thermal exchange with the environment, etc.; E_D, exergy destroyed due to the irreversibilities of the processes; E_g, physical exergy of the gas produced and specified in the installation; E_o, physical exergy of the oil produced and specified in the installation; Others, Physical exergy of water and other contaminants; EF, Installation driving exergy. It can come from an environment external to the installation (e.g., Electricity from the concessionaire) and/or internally (e.g., Natural gas from its own production). In the latter case, the external observer would not see this exergy flow.

5.4.1 Fuel Exergy

As previously identified, the driving exergy of an oil and gas production facility can be obtained by:

- Burning part of the gas produced;
- Burning of other types of fuels external to production, typically diesel oil;
- Electricity from a concessionaire or generated in another installation.

In both situations, this driving exergy will be identified in the analysis as an external input by the symbol (E_F), which also has the contribution of the physical exergy of the oil flow itself (E_{pet}) that reaches the unit.

Strictly speaking, in a facility that uses gas produced in the unit itself as fuel, this is in fact not an external input, but will be considered as such because its chemical exergy consumed to move the plant could be made available to the consumer market. The more efficient the installation, the lower the gas consumption will be and

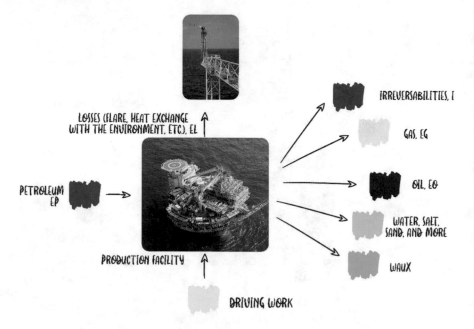

Fig. 5.11 Exergy flows in oil and gas production facilities (Abstraction)

proportionally, a greater quantity of this fuel can be supplied to external customers. This does not generate distortions in the analyses, since for both the incoming and outgoing oil streams (oil, gas, water, and others) chemical exergy is not considered. As already highlighted, only the physical exergy of these currents is considered. The only source of chemical exergy accounted for is fuel gas, in this case, the portion of natural gas produced that is used as fuel.

In reality, as the processes for separating oil in production facilities can be considered as purely physical (the chemical reactions that may occur are neglected), ultimately, the chemical exergy of the fuel is used for a physical separation process. Therefore, it is justified to consider only the physical exergy of the inlet and outlet streams and for the fuel, also account for the chemical exergy (chemical combustion reaction).

Likewise, in the case of installations that use other fuels external to production, such as diesel oil, their chemical exergy must be similarly considered.

5.4.2 Exergy of Products

The main products of a petroleum production facility are the oil and gas outflows.

However, it is not possible to disregard the exergy of water and separated sediments, as well as the exergy used for auxiliary production systems as products, as these are indispensable for production. In the next section, product identification will be defined for two different methods of measuring exergy efficiency.

5.4.3 Exergy Efficiency

For the formal determination of exergy efficiency, it is necessary to identify:

- A product—Represents the desired result, consistent with the purpose of purchasing and using the system or component in question;
- A fuel—Represents the resources (not necessarily fuels in the literal sense) spent to generate the product;
- Both product and fuel are expressed in terms of exergy.

In this context, the exergy balance is written for the steady state as:

$$\dot{E}_F = \dot{E}_P + \dot{E}_D + \dot{E}_L, \tag{5.6}$$

where

E_F Fuel exergy;
E_P Product exergy;
E_D Irreversibilities, or exergy destroyed due to irreversible processes;
E_L Exergy lost or wasted to the environment.

The exergy efficiency of a thermodynamic system is then defined by the expression:

$$\varepsilon = \frac{\dot{E}_P}{\dot{E}_F} = 1 - \frac{\dot{E}_L + \dot{E}_D}{\dot{E}_F}, \tag{5.7}$$

Exergy efficiency aims to indicate the percentage of the fuel's exergy that is found in the product's exergy.

The difference between 100% and ε is due to destruction (irreversibilities generating entropy) and/or loss of exergy in the system. In this section, two ways of calculating the exergy efficiency of the production facility will be analyzed: The Thermodynamic Degree of Perfection and the rational efficiency, defined below.

5.4.4 Degree of Thermodynamic Perfection

The Thermodynamic Degree of Perfection is the most intuitive way to present exergy efficiency. It is basically a relationship between the exergy of the products and the driving exergy of the evaluated process.

It is defined by Szargut et al. (1988g) as the ratio between the exergy of the usable products of the process and the exergy that feeds the process. In our case, this relationship is:

$$\eta = \frac{\left(\dot{E}_o + \dot{E}_g\right)}{\dot{E}_F + \dot{E}_{\text{pet}}}, \tag{5.8}$$

This way of measuring the energy efficiency of a process is interesting when there is a significant difference between the exergy of fuels and that of products.

With this form of calculation, if the total exergy, that is, the physical exergy plus the chemical, is computed for the currents of an oil and gas production facility, the difference between the exergy of the products and that of the fuels will be very small, because the chemical exergy variation of the process is negligible and that of the physics is quite significant. As the magnitude of the chemical exergy of the currents is intensely greater than that of the physical exergy, the relationship becomes insensitive to variations in the process.

5.4.5 Rational Exergy Efficiency

Rational efficiency is indicated for evaluating processes in which what is desired to be measured is not the relationship between the exergy of products and that of fuels, but rather the ratio between the useful, or net, effect in exergy and the exergy used to obtain it. In processes where chemical reactions do not occur, such as the case in question, rational efficiency measures the gain in physical exergy in the process compared to the exergy provided for it.

Rational efficiency is represented for this study by the following equation:

$$\psi = \frac{\text{Useful Effect in Exergy}}{\text{Driving Exergy}}$$
$$= \frac{\left(\dot{E}_{\text{Phys,Oil}} + \dot{E}_{\text{Phys,Gas}} + \dot{E}_{\text{Phys,Water}}\right) - \left(\dot{E}_{\text{Phys,Pet}} + \dot{E}_{\text{Phys,Gás}}\right)}{\dot{E}_F}. \tag{5.9}$$

In an oil and gas production facility, Gas Chemical Exergy ($\dot{E}_F$) is used for a physical Product Separation/Specification process.

However, this way of measuring exergy efficiency does not allow the exergy of the outgoing streams (specified oil, specified gas, and produced water) to be lower than that of the incoming streams (oil and natural gas). In this case, efficiency would be negative, which would make it devoid of physical meaning.

Another problem is that a flow that moves the process can, simultaneously, also take part in the numerator of the relationship, as is the case with the physical exergy of the oil stream that arrives at the installation. It is a fuel for the process, due to its arrival enthalpy, and is part of the numerator of the equation to identify the useful effect. Although Szargut et al. (1988) state that it is "impossible to determine the rational efficiency in this case," the formulation proposed here will identify the physical exergy of the oil stream that only reaches the numerator, as the chemical exergy of the fuel gas in the denominator is much higher than it and would suffer little its influence.

5.5 Management Policies and Guidelines for an Energy Efficiency Program in Oil Production Facilities

This subsection aims to serve as a technical-managerial guide for establishing macro policies for energy efficiency in production facilities, whether in the project phase or in the operational phase.

Below is a list of suggested guidelines for an energy efficiency program in an oil production facility:

(I) Energy Integration: In new projects and in modernization and expansion of existing installations, energy integration (electricity, natural gas, and other utilities) must be sought with the installation itself, with other adjacent installations, with other units or business areas of the company, or even with external agents. For this integration, the following actions must be performed:

   i. Raise opportunities for energy integration of the facility with the company's other business areas and with the existing national structure;

   ii. Evaluate, in all new projects, the potential for energy integration with other facilities or with the existing structure.

   iii. The existence of reservoirs in the concession for gas storage purposes must be verified, in cases of lack of demand for it or the necessary infrastructure for moving it for consumption. This fact allows oil to be produced without the gas associated with it being wasted by flaring or venting.

A typical case of internal energy integration to the production facility is related to the association between its own sources and demands. As this aspect presents itself as a specific and special issue in oil production facilities.

(II) Energy Efficiency in New Projects:

New projects for oil production facilities must be designed in order to seek the best energy efficiency in all their phases. For this, the following actions must be performed:

   i. Use specific design guidelines adjusted to the concept of energy efficiency;

   ii. Implement energy efficiency indicators in new projects;

   iii. Adopt practices for evaluating and comparing the energy efficiency of new projects;

iv. Establish energy efficiency goals to be pursued in new projects based on the proposed indicators.

(III) Energy Efficiency in Existing Installations:

Oil production facilities must review their production processes, adapting them to the concept of maximum energy efficiency whenever technically and economically feasible. As a general rule, installations in operation must:

   i. Apply energy efficiency criteria that reduce operating costs, offer more products to the market and are simple to implement;
  ii. Evaluate its energy efficiency;
 iii. Monitor its energy efficiency using specific indicators; iv. Establish energy efficiency goals to be pursued.

As an example of monitoring and comparison metrics to be applied, both in new projects and in existing installations, we can mention.

(a) Energy Efficiency of the Facilities:

$$\varepsilon = \frac{\text{Energy of products}}{\text{Energy of fuels}}$$

More recently, the concept of exergy has been used to assess the performance of offshore production platforms, replacing energy efficiency based simply on the first law of thermodynamics. We suggest reading Chaps. 2 and 4.

(b) Energy Intensity Index (EII):

$$\text{EII} = \frac{\text{Used energy (GJ)}}{\text{Hydrocarbon production de HC (tons)}}$$

Energy intensity is the ratio between the energy consumed in oil production and the mass of hydrocarbons produced.

The used energy must be expressed in GJ and the hydrocarbon production in tons. Hydrocarbon production includes the production of oil, gas, NGL, and condensate.

This is the indicator suggested by the OGP (International Association of Oil and Gas Producers) for monitoring energy management in oil-producing companies. It is used to replace the autophagy indicator (hydrocarbon production in barrels of oil equivalent/energy consumption converted into barrels of oil equivalent) that was usually used to monitor the energy performance of oil producing facilities (as we can see, autophagy represents an inverse fraction to the Energy Intensity Index). Although it is the indicator indicated by the OGP, the energy intensity index is not able to assess energy efficiency effectively.

Exemplifying this statement, an installation that processes oil with 80% water is impaired in the comparison via the Energy Intensity Index with another that handles oil with 20% water, due to the greater energy requirement of the first one to match the products. By the same reasoning, a unit that has a high energy consumption of auxiliary systems, due to the characteristics of its products such as high sulfur concentration and process complexity, would be depreciated in relation to another simpler plant. The same analysis can be done for installations that perform water injection. That is, this indicator should not be used for comparison purposes, as it depreciates more complex installations. In fact, its use is recommended for monitoring trends and forecasting actions to reduce them.

(c) Associated Gas Utilization Index (AGUI)

$$AGUI = \frac{\text{Used gas}}{\text{Produced gas}}$$

The AGU is in fact not an indicator directly related to the energy efficiency of oil production facilities. However, indirectly, it is a great indicator to be evaluated, especially in oil production facilities that use associated gas as an energy source for their production processes.

From the point of view of the semantics of the word efficiency, this indicator is associated with the efficient use of gas in the production processes of production facilities in general. In an ideal process, its value would be 1 or 100%, depending on the base, in order to demonstrate that all the gas associated with oil production is used in some way, preferably being exported to the consumer market, or used in the production process, usually as the main source of energy (fuel) for the installations.

(d) Thermodynamic Efficiency of Flow Machines

$$\eta_{th} = \frac{w_{ideal}}{w_{real}}$$

The performance of flow machines does not generally assess the efficiency of oil production facilities; however, as this kind of equipment are usually the biggest consumers of energy, it is strongly recommended that this indicator be monitored for the main and most energy-intensive systems in the productive facilities. Monitoring this indicator, in addition to providing verification and adjustments to the operating point of the equipment, its involution may indicate the need for maintenance or overhaul of the machines.

(IV) Renewable Energy Sources

Renewable energy sources should be evaluated as alternatives for installation. They should be adopted whenever there is a demonstration of technical and economic feasibility or benefits in the life cycle cost of the facilities. In order to verify the effective application of renewable energy sources in the installations, the following must be done:

i. Evaluate the potential application of renewable energy sources in existing projects and installations;
ii. Replace expensive and inefficient energy sources with renewable energy sources, whenever technically and economically feasible.

(V) Continuous Improvement Process: For any systematic to preserve its results in the long term, it must be constantly fed back through a process of continuous improvement, in order to make it sustainable. For this, it is suggested that the following actions be carried out:

   i.   Update the energy efficiency policy, guidelines, and targets, periodically adjusting them to the facility's strategic plan;
   ii. Implement an energy efficiency management assessment program aimed at its constant improvement;
   iii. Implement action plans based on the results of these assessments, aimed at preventing or correcting any deviations.

5.5.1 Exergy Efficiency Indicators by the Thermodynamic Degree of Perfection Method

Initially, it is necessary to identify the exergy balance equation:

$$\text{Exergy of inputs} = \text{exergy of outputs} \qquad (5.10)$$

$$\text{Oil} + \text{fuel} = (\text{main products}) + (\text{auxiliary systems}) + (\text{irreversibilities} + \text{losses}) \qquad (5.11)$$

$$\dot{E}_{\text{pet}} + \dot{E}_F = \left(\dot{E}_o + \dot{E}_g\right) + \left(\dot{E}_{\text{others}} + \dot{W}_{\text{Aux}}\right) + \left(\dot{E}_L + \dot{E}_D\right) \qquad (5.12)$$

The equations from (5.10) to (5.12) represent only a purely mathematical approach, since strictly speaking the exergy, as seen in Chap. 2, is not conserved. From a mathematical point of view, the aforementioned balance is achieved by encompassing the destroyed exergy, a portion that identifies the non-conservative nature of the exergy. Physically, the destroyed exergy is not a flow that crosses the control volume; however the methodology demonstrated in the aforementioned equations can be used for calculation purposes.

Considering all the unit's energy consumption as products, the exergy efficiency of an oil and gas production facility could be:

$$\varepsilon = \frac{\left(\dot{E}_o + \dot{E}_g\right) + \left(\dot{E}_{\text{others}} + \dot{W}_{\text{Aux}}\right)}{\dot{E}_F + \dot{E}_{\text{pet}}}. \qquad (5.13)$$

Although the exergy of water and other contaminants, as well as that of auxiliary systems, do not actually represent the products of the installation, they must be identified as such so that comparisons between different installations are not prejudiced. Exemplifying this statement, a facility that processes oil with 80% water would be disadvantaged compared to another that handles oil with 20% if the latter's exergy was not counted as a product. By the same reasoning, a unit that has a high energy consumption of auxiliary systems, due to the characteristics of its products such as high sulfur concentration and complexity of processes, would be depreciated in relation to another simpler plant.

However, the exergy efficiency indicator that will be used to determine the exergy efficiency of the installation will only consider the products [specified oil, specified gas, produced water ($\dot{E}_{PW}$) and injection water ($\dot{E}_{IW}$)], according to Eq. (5.14), presenting the exergy of the auxiliary systems at a specific indicator identified in Eq. (5.15).

$$\varepsilon_{\text{Main}} = \frac{\left(\dot{E}_o + \dot{E}_g\right) + \left(\dot{E}_{PW} + \dot{E}_{IW}\right)}{\dot{E}_F + \dot{E}_{\text{pet}}}. \tag{5.14}$$

Other indicators can be obtained to identify the weight of each element in the exergy consumption of the process, as follows:

$$\chi_{\text{Aux}} = \frac{\dot{W}_{\text{Aux}}}{\dot{E}_F + \dot{E}_{\text{pet}}}, \tag{5.15}$$

which evaluates the contribution of auxiliary production processes (which do not transfer exergy to the products) to the exergy consumption of the installation. A high value of this factor represents inefficiency of auxiliary processes. To optimize the exergy efficiency of production facilities, it is important to minimize these consumptions.

Identically, to assess losses we have:

$$\chi_{E_{L+D}} = \frac{\left(\dot{E}_L + \dot{E}_D\right)}{\dot{E}_F + \dot{E}_{\text{pet}}}, \tag{5.16}$$

which identifies the contribution of losses, mainly due to Flare burning, to the exergy consumption of the installation ($\dot{E}_L$). Minimizing these losses is essential. This indicator also accounts for the effect of irreversibilities ($\dot{E}_D$), responsible for the generation of entropy and consequent exergy destruction in the processes carried out in the installation, both for the main products and for others.

In real installations, idealized reversible processes do not exist. In any case, investment must be made in more efficient processes to reduce the effect of irreversibilities in exergy destruction and thus the unit's productive capacity.

In this indicator, it is not possible to separate the lost from the destroyed exergy. For separation to occur, it would be necessary to quantify the losses due to Flare, machine exhaust, thermal exchange with the environment, among others. This procedure is not possible using the simplified methodology proposed here. In reality, the calculation of lost plus destroyed exergy is done according to the balance in Eq. (5.17) below:

$$\left(\dot{E}_L + \dot{E}_D\right) = \left(\dot{E}_F + \dot{E}_{pet}\right) - \left[\left(\dot{E}_o + \dot{E}_g\right) + \left(\dot{E}_{PW} + \dot{E}_{IW}\right) + \left(\dot{W}_{Aux}\right)\right]. \tag{5.17}$$

5.5.2 Exergy Efficiency Indicators Using the Rational Efficiency Method

In the same way as in the previous section, at first it is necessary to identify the exergy balance equation:

$$\text{Fuel exergy} = (\text{useful exergy effect}) + (\text{losses} + \text{destruction}), \qquad (5.18)$$

$$\text{Fuel} = (\text{main products} - \text{oil}) + (\text{auxiliary systems}) + (\text{losses} + \text{destruction}), \qquad (5.19)$$

$$\dot{E}_F = \left[\left(\dot{E}_o + \dot{E}_g\right) + \left(\dot{E}_{\text{others}} + \dot{W}_{\text{Aux}}\right)\right] - \left[\dot{E}_P\right] + \left(\dot{E}_L + \dot{E}_D\right), \qquad (5.20)$$

Thus, if all energy consumption of the unit were considered as a useful effect, the exergy efficiency of an oil and gas production facility could be:

$$\varepsilon = \frac{\left[\left(\dot{E}_o + \dot{E}_g\right) + \left(\dot{E}_{\text{others}} + \dot{W}_{\text{Aux}}\right)\right] - \left[\dot{E}_P\right]}{\dot{E}_F}, \qquad (5.21)$$

However, analogously to what was developed in the previous section, the exergy efficiency indicator to be used in this study will be the one represented in Eq. (5.22).

$$\varepsilon_{\text{Main}} = \frac{\left[\left(\dot{E}_o + \dot{E}_g\right) + \dot{E}_{\text{others}}\right] - \left[\dot{E}_P\right]}{\dot{E}_F}, \qquad (5.22)$$

where $\dot{E}_{\text{others}}$ represents, in addition to the water stream and sediment produced, also the injection water. This indicator presents a problem that has already been identified in Sect. 5.4.5. It can be negative when the magnitude of the exergy of the products is small compared to the exergy that enters with the oil which is a rare situation.

For this reason, the other indicators obtained to evaluate the weight of each element in the exergy consumption of the process, analogous to those identified in the previous section, could have no physical meaning when using the rational efficiency method.

As a way of solving this inconvenience, the weights of each element for exergy consumption will be determined according to the relationship between each one of them and the fuel exergy. In this way, they will indicate how much of the fuel's exergy each element of the analysis consumes.

$$\chi_{\text{Aux}} = \frac{w_{\text{Aux}}}{\dot{E}_F}, \qquad (5.23)$$

The indicator in Eq. (5.23) determines the contribution of auxiliary production processes to the exergy consumption of the installation. A high value of this factor represents inefficiency of these processes. To optimize the exergy efficiency of production facilities, it is important to minimize this consumption. Identically, to evaluate losses and exergy destruction, we have:

$$\chi_{E_{L+D}} = \frac{\left(\dot{E}_L + \dot{E}_D\right)}{\dot{E}_F},$$
(5.24)

which determines the contribution of losses and the effect of irreversibilities (responsible for the generation of entropy) to the exergy consumption of the installation. Losses occur mainly in Flare burning and in the exhaust of thermal machines ($\dot{E}_L$). Minimizing these losses is interesting for reducing fuel consumption and consequently increasing the plant's energy efficiency.

The destroyed exergy ($\dot{E}_D$) accounts for the effect of irreversibilities, responsible for the generation of entropy and consequent exergy destruction in the processes carried out in the installation, both for the main products and for others.

5.6 Good Energy Efficiency Practices in Production Facilities

In this section, important information will be provided to consider in the design, evaluation, and operation of production facilities.

i. Mass and Energy Balance

   The mass (or material) and energy balance is a simple mandatory study in the design of installations in the chemical, petrochemical, and energy industries. It is widely used in these industries with the purpose of evaluating processes and sizing equipment; however, it is normally not used to identify opportunities for reducing consumption or energy degradation.

ii. Evaluation of Energy Losses

   In addition to the assessment of energy losses mentioned, the following are worth highlighting specifically with regard to their application in oil production installations:

   (a) losses of hydrocarbon vapors in tanks, including the tanking of crude oil from FPSO and FSO;
   (b) gas losses in atmospheric vents.

iii. Identification of Best Practices and References of Technological Excellence

   As technology is always in constant evolution, we must carry out studies aimed at identifying best practices and references of technological excellence that can be applied to industrial projects (new, or in renovations or expansions of existing facilities), according to the following steps:

   (a) research more current applications in similar projects;

(b) determine better options in terms of energy efficiency;

(c) evaluate the cost/benefit ratio and technological maturity;

(d) research best practices internal and external to the company or installation for each phase of the project, identifying references of technological excellence (benchmark);

(e) research the use of renewable and alternative energies.

In addition, design practices or the identification of actions to improve the energy efficiency of existing installations must consider the operating cost of energy through the study of the installation's life cycle. Thus, operating costs arising from differences in energy efficiency of the unit's main dynamic, thermal, and electrical equipment must be considered in the process of selecting and purchasing them, through an economic analysis over the expected operational life of the facilities.

iv. Optimization of Thermal Exchange Processes

Thermal exchanges represent important processes in the production and specification of oil and other hydrocarbons, in addition to presenting themselves as excellent opportunities for energy integration and use. In this way, we must always:

- Check the energy integration potential as a way to reduce the demand for hot utilities (fuels, steam and intermediate thermal fluids) and cold utilities (air, cooling water, cooling fluids).
- Optimize the arrangement, ordering, and dimensions of heat exchangers using analysis tools such as pinch technology.
- Prioritize direct heat exchange between the currents involved, adopting the use of intermediate thermal fluid only when there are limitations imposed by distance or for safety reasons.
- When selecting the thermal fluid, among the substances that meet the project's technical and safety requirements, choose the one that provides the greatest heat transfer per unit mass, minimizing pumping operations.
- Specify the type of heat exchanger that presents the least energy efficiency degradation, considering the desired campaign and the fluids involved, their properties and pressure drop limitations.
- Note that, in addition to the most commonly used technological solutions such as shell and tube heat exchangers, plate exchangers and regenerative rotary heaters, there are alternative technologies. Depending on the type of application, options such as compact heat exchangers and heat tube heat exchangers or thermosyphons can be used, aiming at increasing energy efficiency.
- Provide for measurement and transmission of temperature and pressure data before and after passing through each heat exchanger of a battery. The measurement of flow in the limits of this battery must also be foreseen. This allows the monitoring of the deposition, including allowing the use of specific computer programs. In cases where such measures are not possible, at least measuring instruments should be provided to allow the evaluation

of the efficiency degradation of the heat exchangers by closing the balance of the exchanger bank.

- In marine installations, the possibility of capturing sea water for the cooling circuit at greater depths must be evaluated in order to obtain lower temperatures (less than 15 °C), reduce the flow rate and the size of the heat exchangers.

v. Analysis of Use of Mechanical Energy of Currents

The flows that cross oil production facilities generally have high mechanical energy (high pressure and temperature). In the previous item, we dealt with the thermal exchanges and exploitation of the temperature of these flows; however, the exploitation of the pressure inherent to them is not very common and isenthalpic expansion processes are trivial, for these reasons we must:

- Identify current pressure reductions that may have their mechanical energy harnessed. By using, for example, turboexpanders, reciprocating expanders or steam and hydraulic turbines instead of expansion valves, we can achieve electricity generation or mechanical drive. In all cases, it is recommended to investigate the potential use of cooling provided by expansion of the fluid for use in another part of the process, in auxiliary systems, utilities or even in liquefaction processes.
- The possibility of using low pressure currents as driving energy for ejector systems for moving fluids must be verified.

vi. Analysis of Use of Disposal Streams
We constantly see production facilities that discharge streams in high physical imbalance with the environment. These currents can represent excellent opportunities to avoid wasting energy. Therefore, whenever possible, we should:

- Identify, in the project, all environmental disposal streams that have energy potential to be used, such as fuel gases directed to the torch, combustion gas emissions, purge or relief systems, and other non-energy uses of gas, including FPSO and FSO tanking and adopting solutions to take advantage of these currents.
- Install heat exchangers for combustion gases.
- Maximize the use of the gas that would be ventilated or burned in the torch through, for example, a compression and treatment system to return it to the process, or its use as fuel for utilities. Torch burning should only take place in operational emergency conditions.
- The use of low-pressure atmospheric ventilation of natural gas should be avoided, as this gas is basically composed of methane, which has a global warming potential far greater than CO_2.

vii. Inclusion of Essential Measurement Points for Energy Efficiency Analysis

Currently, with the growing demands for improving energy efficiency in industrial plants, the professional specialist in this subject is sent to the field to identify opportunities for advancement, however, he faces a very peculiar difficulty: there is no way to calculate the level of general efficiency or of isolated plant systems, as there are no essential measures to perform the calculation. In this way, something that is not measured, does not allow the identification of the current status and, consequently, does not allow the effective implementation of improvement actions, which, if used, their gains will never be fully or precisely quantified. Thus, the first step to increase the energy efficiency of an equipment, system, or plant is to provide the necessary measurements for its calculation. At this point, we invite the reader to return to Chap. 6, referring to Measurement, which presents the necessary measurements for each type of equipment or system. New projects or modifications to existing installations must consider the following aspects:

- The choice of the type of variable and the places where the measurements will be carried out, with a focus on the energy efficiency of the project, must be such as to allow the evaluation of energy consumption and its operational cost. The above concept should be applicable to equipment, process units, sets of units, regions, or areas of the plant. The objective is to make it possible, by direct measurement or by closing the balance sheet, to quickly identify the loss and locate it.
- The measured variables must be available in the supervisory system for continuous monitoring, performing calculations and triggering alarms, in more critical cases.
- The grouping of currents with a single measurement should be avoided, a fact that masks and makes it difficult to identify losses. Measurement should preferably be applied to individual consumers.
- In the absence of direct measurement in the "vent" lines, indirect measurement should be used through correlations established by equipment suppliers, based on their measured parameters, or through the use of thermodynamic simulators, to estimate the ventilated flow.

viii. Optimization of Fluid Handling Systems

In oil production facilities, the equipment responsible for moving fluids, mainly the gas compression and water injection systems, represent the greatest demand for mechanical energy (usually electrically driven, or less frequently by steam) of the unit. In this way, actions with a positive impact on the energy efficiency of these systems should be widely used. Some of them are highlighted below:

- The design of ducts and piping must have as one of the assumptions the minimization of load losses caused by curves, accessories, roughness of the tubes, among others. In piping design, the energy cost resulting from head losses during the life cycle of the duct system must be confronted with the cost of acquiring pipe material, aiming at the optimized specification of its diameter.

- Pumps, compressors, and other equipment in fluid handling systems must be dimensioned in order to operate in the operational range of best energy efficiency (We call the reader's attention again to questions related to the term operational range. An oil-producing field has a production curve, where the current flows of oil, gas and water produced vary over time, meaning that, in general, the equipment does not operate at a fixed point, but along an operational range).
- Where there is a need to heat the product before pumping, consideration should be given to using leftover energy from the unit's processes (turbine exhaust gases, low pressure steam, etc.) and, in their absence, renewable energy sources (such as to solar).

ix. Thermal Sources and Demands

- Energy integration between the different process streams must always be sought in order to take advantage of their thermal potential and reduce the demand for hot and cold utilities (fuels, steam, thermal fluids, cooling water, etc.). The energy integration chosen must be the one that provides the greatest overall energy gain for the plant under analysis.
- Temperature differences between hot and cold sources and pressure drops in heat exchangers are sources of entropy generation and must be minimized. Therefore, when specifying and purchasing a heat exchanger, the energy efficiency degradation effect must be observed for the constructive types under analysis, considering the desired campaign, the fluids involved and the head loss.
- It should be analyzed whether the application for which the project is intended allows the use of technological solutions of greater efficiency without compromising reliability, such as compact heat exchangers and heat tube exchangers, as an alternative to technologies in common use (hull and tubes, plate exchangers, and regenerative rotary heaters).
- The option for associating exchangers in series or in parallel must consider, in addition to cost aspects, the comparison between the efficiency gain in heat transfer in series association and the energy cost of pumping due to its greater pressure drop in compared to parallel association.
- The possibility of using the thermal energy of the natural gas acquired in the compression process as a source of heating must be evaluated.

x. Oil Separation and Treatment Processes

These processes are the main thermal energy demanders of an oil production facility. Mainly in the production of heavier oils (low API grade), this energy demand is very significant, a fact that makes it even more important to take advantage of the thermal sources available in the installation. Another issue is related to the reduction of the thermal load, which can be obtained with the correct destination of the oil and water streams. These and other aspects are identified and detailed below:

- The oil separation and treatment processes must be dimensioned in order to minimize the thermal load for the product to conform to the required specifications. Sending free water to the separation system should be as little as possible.
- Liquid flows to the slop vessel (waste vessel) must be minimized, prioritizing the sending of dominant oil flows at low pressure (high pressure flows go directly to the production separators). Three-phase separation must be provided for the oily waste vessel.
- For cases of units with electric and thermal energy generation system through turbogenerators, the configuration of the oil processing plant must be chosen in order to foresee the use of the heat made available by the exhaust gases of the turbogenerators. The heating temperature of the thermal fluid must be optimized to obtain the maximum thermal use of the hot gases exhausted by the turbogenerators.
- The installation of furnaces to meet the thermal demand must be avoided. If the installation of furnaces is indispensable for supplementing the necessary heat, the use of natural gas as fuel should be preferred, to the detriment of the use of diesel oil or crude oil, in order to reduce atmospheric emissions to the essential minimum.
- In cases where the thermal heating demand required by the installation is higher than the heating heat available in the WHRU such as, for example, in the case of heavy oil fields, with high salinity of the formation water and low gas–liquid ratio (GLR), alternatives that enable energy integration with other facilities, or the use of oil processing plants that make it possible to reduce the required thermal demand, or the use of other thermal sources existing in the facility itself, should be prioritized.
- The use of heat exchangers to take advantage of the treated oil, by cooling the oil to be stored or transferred, should be adopted as an alternative to reduce the thermal demand required to heat the flow produced.
- For cases in which the temperature or the discharged water flow in the electrostatic treaters is high, resulting in a great loss of heat with the discharge of hot water, the plate heat exchangers must be configured to work as oil heat exchangers or as water heaters. An arrangement of valves must allow the replacement of the hot fluid fed from oil to water, in each set of plates, according to the production curve. During the decline in oil production and increase in water production, some sets of plates intended for harnessing oil heat should be allocated for harnessing water heat.
- Processes that have a low demand for electrical energy and that minimize the consumption of thermal energy, such as free water separators and washing tanks, must be prioritized. The decision on the selected process must be supported by a global energy efficiency analysis of the unit.
- The use of Washing Tanks technology must be adopted for cases of heavy oils or with high salinity of the formation water, with the aim of reducing the thermal heating demand required. Heat recovery and production heaters must be installed downstream of the Wash Tanks.

- Opening pressure relief valves (PSV) must send inventory to the process or to the torch.

xi. Natural Gas Specification and Treatment Processes

Commonly, the pressure of some gas streams is raised to higher pressure levels than necessary for use, which promotes greater energy consumption in the production facility. In addition, the gas is often specified in the installation for the external consumption specification standard, which is not mandatory for internal uses and promotes higher energy consumption. These points are detailed below:

- The natural gas specification and treatment process must be just what is necessary to specify the gas (pressure, temperature, and maximum allowed contaminant content) according to its respective uses: export, reinjection, "gas lift" or internal consumption, of in order to demand the least amount of energy from the installation.
- Degradation of gas pressure levels in this system must be avoided.
- Processes that require high pressure losses and consequent recompression must be avoided.
- An energy efficiency and gas specification analysis must be carried out for the purpose of selecting the separation technology to be used (chemical absorption and membranes). The following aspects must be taken into account:

  - the highest thermal demand in chemical absorption process through amine.
  - the amine chemical absorption process is the most suitable for the condition of low CO_2 partial pressure in the feed and in the final product (treated gas);
  - In the condition of high CO_2 partial pressure in the feed and in the final product (treated gas), the use of physical processes (permeation with membranes) is the most appropriate;
  - The use of a hybrid process, mixing the two technologies, should be evaluated.

- In cases where the gas produced has a relatively high amount of CO_2 (above the value established for export), the use of the available CO_2, separated by the natural gas treatment system, must be evaluated for use in the advanced recovery process of oil (EOR—Enhancing Oil Recovery), or its supply to non-energy consumers of natural gas.

xii. Fuels

The use of fuel supplied to the production facility by means external to it promotes, in addition to possible internal inefficiencies, high energy consumption for transport, and all the logistics that involve them, including incurring greater emissions of greenhouse gases. Thus, the following aspects must be evaluated when selecting the fuels to be used in the production facility:

- In cases where the installation produces natural gas, the use of other fuels such as diesel oil and fuel oil must be avoided, except for security and emergency systems.
- When the use of external fuel is necessary, the indirect costs, such as transportation, handling, and storage, must be considered when choosing it.
- In facilities where natural gas production is not sufficient to meet their demands and where energy integration (gas, electric, and thermal energy) with other units is not feasible, the standardized use of fuels for all applications must be sought existing in the plant.
- The possibility of using hydrated alcohol as fuel, replacing other more carbon-intensive fuels, such as gasoline, diesel oil, and fuel oil, should be evaluated.

xiii. Natural Gas Produced

The natural gas produced is established as the main source of fuel for the oil production facilities, in addition to being used in the processes and utilities of the facilities and being one of the products of the facility, so we must maximize its delivery to the consumer market. Thus, in relation to the gas produced, the following aspects must be considered for its greater use and for the reduction of the degradation of its energetic potential (essentially of pressure degradation):

- Projects for production facilities must be specified in such a way as to promote maximum use of the natural gas produced, preferentially directing it to the consumer market.
- Projects must provide for the possibility of depressurizing all plant equipment and systems that operate with high pressure gas (including gas pipelines) to the fuel gas system or to the existing compression system. This gas should preferably be used as fuel (it can also be used for non-energy uses). For this, the fuel gas system and the suction vessel of the discharge compressors must be specified with the necessary resources (physical and control) for two power supplies (treated and untreated gas).
- The use of atmospheric venting of waste gases from processing systems, treatment, or energy or non-energy use of natural gas should be avoided.
- The possibility of different specification of natural gas with a high content of contaminants (CO_2 and H_2S) for non-energy and energy uses must be identified, such as, for example, fuel for motive flow machines (gas turbines and internal combustion engines).
- The gas compression system must be specified to send the condensate generated in the second and third stages directly to the oil-processing plant. The condensate generated in the first stage (essentially water) must be sent directly to the water treatment plant.
- The low pressure gas coming from the drainage system must be sent to the suction of the Vapor Recovery Units (VRUs). If the gas pressure level is not

compatible with the suction of the low compressors, such compatibility must be carried out through ejectors, having natural gas or condensate generated by the compression system as the driving force.

- For depleted gas reservoirs, the possibility of their reuse for gas storage purposes must be evaluated, taking advantage of the available surface facilities and thus reducing the burning of gas in them.

Additionally, the following items must be considered respectively for energy and non-energy use of the natural gas produced.

xiv. Energy Use of Produced Natural Gas

- The natural gas produced must be used energetically for motive flow machines. The use of more efficient machines should be sought in order to maximize the transfer of gas as a product to the consumer market.
- This maximization, however, must not result in the replacement of natural gas by another fuel that is more carbon intensive (such as diesel oil, fuel oil, and crude oil) as the preferred fuel for the generation of energy to be consumed at the facility.
- The extraction of gas from the compression system for this consumption must be done as close as possible to the pressure level required by the machines. The possibility of extracting the gas at intermediate points in the compressor frame at a pressure level suitable for each demand must be evaluated.
- In all fuel gas consumption in which there is a significant difference in pressure levels in relation to the supply source, the use of ejectors in place of control valves must be evaluated. In this way, it is possible to recover the gas that would be ventilated at lower pressure levels by the process plant, sending it to low-pressure gas recovery units (VRU).
- In cases of production from gas wells (gas or high Gas/liquid ratio reservoirs), the possibility of using it directly by the fuel gas system or by the gas dehydration unit must be checked.

xv. Non-Energy Use of Produced Natural Gas

- Non-energy use of produced natural gas must be avoided. Processes that do not require natural gas for non-energy purposes should be prioritized, such as: vacuum flotation, vacuum deaeration, regeneration of triethylene glycol with other fluids, such as nitrogen (e.g., sealing process plant vessels with nitrogen, among others).
- In cases where the non-energy use of gas is unavoidable, all its effluent must be sent to steam recovery units (VRU). In situations where the pressure level is insufficient to direct the gas directly to the VRU, the use of ejectors must be evaluated in order to promote the necessary pressure increase so that it is sucked by the VRU compressors or used directly by the consumers of gas at low pressure, avoiding its ventilation to the atmosphere.

xvi. Process Optimization or Adjustments

Although it seems repetitive, it is very important for the reader to understand that oil production facilities have certain very particular characteristics and in this chapter we have insistently drawn attention to them.

The first important aspect is related to the fact that we do not know exactly, in the design and construction phase of the production facility (the moment in which the equipment is specified and purchased), what will be the characteristics of the fluids produced by it. In other words, there is only an estimate with variable accuracy of the composition of the fluids, the BSW (Basic Sediment and Water), the Gas/Liquid Ratio, the pressure and temperature of the formation, among other characteristics. We use to say, ironically, that we only know that the installation will operate in different conditions than what we foresee in the project.

Another aspect is related to the oil, gas, and water production curve. The flows produced by these streams vary significantly over the life of the producing field. In addition, it is very common to have changes in the characteristics of the fluids produced.

All of this makes the task of designing an oil production facility efficiently over its entire life quite challenging. In general, the project is carried out based on some points of the production curve: maximum production of oil production, maximum production of gas, maximum production of water, maximum demand for electrical energy, maximum demand for thermal energy. These maximum points make it possible for systems and equipment to be specified in order to meet these demands. However, mainly due to the aforementioned uncertainties, the points of minimum or intermediate demands are generally not studied in depth in the design phase.

All these issues make the task of operating a production facility efficiently at all points in its lifecycle an almost impractical task. In this way, the only adequate way to deal with the uncertainties and production variations of a producing field in order to obtain excellence in energy efficiency is to carry out adjustments or optimizations in the production processes. In practice, we need to make these adjustments from time to time. Here are some examples of tuning and optimizing actions:

- The adjustments of the plant's control system related to the set points of level, temperature, pressure, among others, must be checked and adapted constantly in order to keep them at values that provide the optimization of the energy and productive efficiency of the installation. This task must be carried out through computer simulations of the process.
- Control valves must be evaluated to ensure that their operating ranges are adequate for the new operating conditions. Eventually, they must be replaced for the control to be effective in the optimal situation.
- Flow machines must be constantly evaluated in order to always keep them as close as possible to their maximum efficiency points (BEP—Best Efficiency Point). Eventually, under certain operating conditions of the life cycle, these equipment operate so far from the BEP that, in addition to a significant reduction in energy efficiency, there is a drop in reliability, as they operate in a destructive range.

In these cases, this equipment must be replaced by others that present efficient operating ranges that are more appropriate to the new operating conditions.

xvii. Modularization

Another way of dealing with uncertainties and variations in the operating conditions of the process plant is through the practice called modularization. This practice refers to the design of systems in modules. In each phase of operation of each system, a certain number of modules is required. Thus, the modules can be mobilized or demobilized according to the operating condition in effect at the installation. Furthermore, these modules must be specified in such a way as to promote their easy replacement by others that are more adequate to the operational point in force. Next, we identify and detail some aspects related to modularization:

- To enable the use of different energy solutions for the facilities, modular arrangements of systems and equipment should be prioritized to facilitate future mobilizations and demobilizations.
- In the case of energy generation systems, their modularization should even allow the use of different technologies that are more appropriate for each phase of the installation's life cycle.
- The modularization of all heating or cooling equipment and systems in the oil, gas, and water treatment processes must be planned to ensure the best energy efficiency and use of the capacity of the modules in each phase of the unit's life. To meet this requirement, facilities must be provided for the assembly and disassembly of equipment during the life cycle of the installation.

xviii. Energy Conversion

The energy conversion system (popularly identified as the energy generation system) must be specified carefully, as it represents, in general, the main source of irreversibilities in the production facility. Thus, the aspects listed below should be evaluated:

- The selection of the primary machine of the generation system of the production facilities, as well as the respective thermodynamic cycle, must take into account the thermal and electrical demand profile of the same throughout its useful life. The decision for the system to be applied must be supported by a global analysis of the energy efficiency of the unit.
- For units with a steam cycle electrical and thermal energy generation system, this fluid must be prioritized to supply the thermal demand required by the installation. For each thermal demand, steam must be used at appropriate pressure and temperature levels for the applications, in order to minimize energy degradation in the steam system. The steam cycle must be optimized seeking the best overall efficiency of the energy system, through the best relationship between thermal demand and required electrical demand.
- In installations with low thermal demand (in the case of light oils and at high temperatures), a detailed assessment of the use of the combined cycle must be

carried out to increase the efficiency of the electricity generation system. Additionally, the use of existing heat sources in processes and equipment can be directed toward air conditioning in offices, accommodation, and control rooms.

- The possibility of using renewable energy should be evaluated, mainly solar for heating and photovoltaic for powering small loads and others such as wind and wave and current energy.
- The use of potential energy from the discharge of water from the open cooling circuit and produced water must be evaluated for the generation of electricity through hydraulic turbines.

5.7 Current Technological Limitations and Issues for the Future

In this section, we will look at possibilities for increasing energy efficiency in oil production facilities planned for the future. These solutions are not currently put into practice due to technological limitations or the existence of some paradigms that need to be broken before their effective use. Basically, the solutions are related to improving the efficiency of primary energy conversion systems, increasing integration between existing energy sources and demands, making better use of latent energy from producing wells, advancing technology for transferring thermal energy and the intensification of the use of renewable energy sources.

5.7.1 Use of Combined Cycle for the Main Energy Conversion System

In fact, the use of the combined cycle in onshore production facilities should be a priority due to the maturity of the technology and the gains in efficiency, reduction of emissions and operating costs that it promotes. In marine installations, there are still some paradigms related to the initial investment value, weight, occupied space, and the difficulty of operating the system and maintaining the water quality required by it. All these issues have gradually become more fragile due to the practical application of combined cycles in some production facilities in the North Sea and other areas, proving the technical and economic viability of this solution in offshore units. This fact is due to the evolution of conventional boiler designs (with tube) and once trough boilers, making them more compact and simpler to operate. The Figs. 5.12 and 5.13 show applications of combined cycle for offshore production facilities.

The use of a combined cycle, in addition to significantly improving the energy efficiency of energy conversion (typically, the energy efficiency of a simple or open cycle is around 30% and of a combined cycle is greater than 50%), allows for greater operational flexibility and better adjustment to variations in energy demands throughout the installation's life cycle. In moments of greater demand for electricity, we can

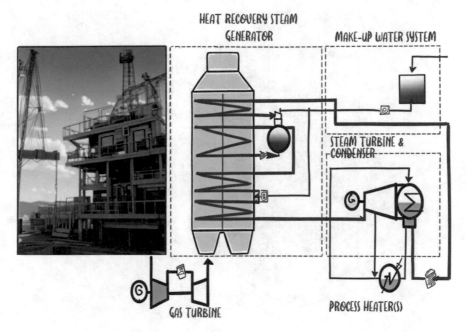

Fig. 5.12 Example of a combined cycle in a maritime installation with a conventional pipe boiler (two gas turbines feed a conventional boiler to drive a steam turbine)

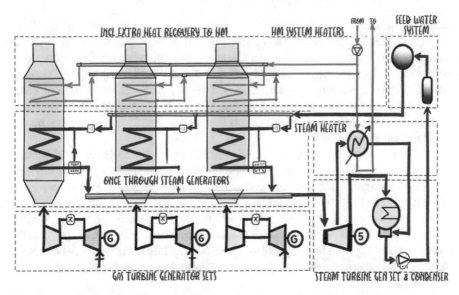

Fig. 5.13 Example of combined cycle in offshore installation with once trough type boilers (3 gas turbines individually feed their boilers to activate a steam turbine)

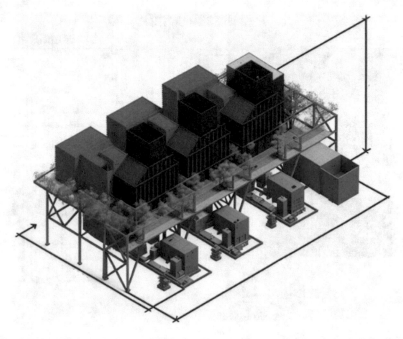

Fig. 5.14 3D model of the arrangement of Fig. 5.14

prioritize the operation of the steam turbine at maximum capacity, by sending all the steam generated in the boilers to it and in situations of increased demand for thermal energy, we can extract steam in intermediate stages turbine or even fresh and super-heated steam from the boiler itself, sending it to the process that requires thermal energy, in order to operate the steam turbine at partial load or even, in extreme cases of thermal need and low electrical demand, turn it off there. Figure 5.14 shows an example for a combined cycle application in a marine installation with once trough compact heat recovery boilers.

In this arrangement, the steam turbine has the capacity to generate electricity compatible with the gas turbine used in this system, as each gas turbine at full load provides 33% more of its mechanical energy generation potential (exergy) in its exhaust with possibility of conversion into mechanical energy in the steam turbine. In other words, each unit of kW (electric kilowatt) generated in the gas turbine corresponds to 0.33 kWe generated in the steam turbine. It is worth noting that the thermal energy required for the process is also obtained from the recovery boilers, by means of extra thermal exchange sections positioned in the low temperature region of the exhaust gases.

Figure 5.15 shows the 3D model of the previous arrangement, demonstrating that the space occupied by the once trough boiler is compatible with that occupied by commonly used conventional WHRU, reinforcing that this solution is quite competitive for application in marine units.

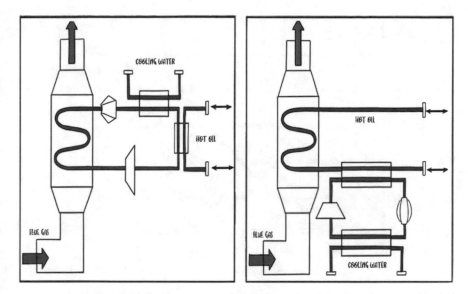

Fig. 5.15 Electric energy production and heat use in combination

5.7.2 Intensification and Evolution of Use Cogeneration Cycles

As in the previous subsection, the most important source for cogeneration in a production facility is the exhaust from thermal machines such as the gas turbine. However, cogeneration is a very common application solution in oil production facilities. Thus, in this subsection we intend to open the reader's mind to the various possible configurations for heating the loads to be processed, in addition to exploring current evaluations on cycles and working fluids in evaluation and research. To increase the efficiency of cycles combined with main cogeneration, current research follows the line of intensifying the use of available energy in primary sources through cycles of the bottoming cycle type with the use of conventional, organic, or alternative working fluids.

Figure 5.16 presents two options for combined cycle with cogeneration with possible application in oil production facilities, mainly with the use of alternative working fluids.

The use of CO_2 or organic fluids such as hydrocarbons and those of the HFC type as a working fluid to replace water/steam has been intensively studied. In addition, there is research into using air or nitrogen in Brayton cycles.

Direct heat exchange with the exhaust of gas turbines can always be considered, mainly by choosing the appropriate working fluid. In addition, two other solutions can be evaluated for heat recovery from exhaust gases:

Figure 5.16a presents an example of direct heating of the working fluid by the gas turbine exhaust gas and heat exchange downstream of a turbine with exhaust

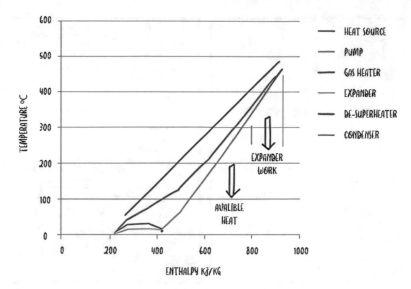

Fig. 5.16 Heat available in a bottoming cycle of a stage with CO_2 as working fluid

temperature hot enough to heat the process. This example is perfectly suited to the use of supercritical CO_2 as a working fluid, due to its high degree of superheating at the turbine outlet (allowing the use of this heat in the process).

Figure 5.16b shows another example for use in heavier oil production facilities with high heating demand. This is a perfect example of a bottoming type cycle (where the thermal demand is the main one), where a thermal oil is heated to supply the thermal demands of the process and an alternative thermal fluid (hydrocarbon, HFC, CO_2, nitrogen or air, as some examples) is used in a secondary circuit to generate additional mechanical work or electrical energy. In times of high thermal demand, the secondary circuit can be disabled and all the thermal energy would be used in the process.

The use of bottoming-type cycles with alternative working fluids such as CO_2 will be an important future solution for taking advantage of low-enthalpy energy sources, as we will see in the next subsection.

The use of working fluids in a supercritical state provides an additional advantage, since there is no phase change during the heat exchange process and, in this way, the thermal exchange surface between the hot and cold sources is increased, as shown in Fig. 5.17. In addition, with the absence of phase change, the variation in the specific volume of the fluid is small, facilitating its handling.

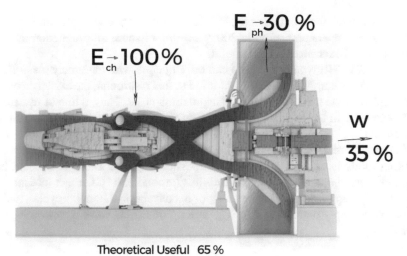

Fig. 5.17 Exergy flow in a gas turbine

5.7.3 Energy Recovery from Heated Gas in Compression Systems

The produced gas is compressed in the production plant to high pressure levels (100–450 bar) for transport, reinjection or for gas lift. Compression produces a lot of heat and compressed natural gas can reach high temperatures, above 150 °C after each stage of compression, configuring an interesting thermal source to be used. However, it is not possible to use thermal cycles with conventional working fluids due to the insufficient temperature for a conventional Rankine cycle generating superheated steam. In this way, the application of thermodynamic cycles such as that of the secondary circuit of Fig. 5.16b of the previous subsection (with the use, for example, of organic fluids or CO_2 as working fluid) is configured as the most adequate way to convert this source of thermal energy into useful mechanical work.

Thus, the heat of the compressed gas (at 100–180 °C) must be recovered in a WHRU (Waste Heat Recovery Unit) type exchanger, where the working fluid can be evaporated and, preferably, superheated before generating mechanical work or electrical energy in a turboexpander. In this way, the WHRU replaces the inter-cooler (between stages of compression) or the aftercooler after the completion of gas compression.

Energy recovery is moderate and drastically depends on the compressed gas temperature, in addition to space constraints for the exchanger and its current state of the art. Furthermore, the choice of the working fluid according to its characteristics and the temperature levels of the hot and cold sources is important for the best use of the energy available in this source.

One of the main challenges for taking advantage of this source is the design of compact heat exchangers, such as printed circuit heat exchangers (PCHE—Printed

Circuit Heat Exchangers). Currently, PCHE already operate with pressures of up to 600 Bar and temperatures of up to 200 °C. Some offshore oil production facilities use this type of exchanger for gas cooling.

However, PCHE type exchangers can only operate with extremely clean fluids, due to their very narrow channels. In addition to this restriction, the existence of only one supplier in the market causes high acquisition costs. Thus, a line of research to be developed will be the study of other types of compact, efficient and robust heat exchangers.

Another thermal source that can be used to feed these cycles with alternative working fluids is the exhaust gas of combustion engines. These equipment have lower exhaust temperatures than gas turbines (around 350 °C), a fact that makes it difficult to generate superheated water vapor in conventional Rankine cycles, making it feasible to apply organic fluids, CO_2, among others, as work in bottoming-type cycles.

5.7.4 Recovery of Natural Energy from Oil- and Gas-Producing Wells

The processes of separation, stabilization, and specification of products in an oil production facility need to expand the produced flow, reducing its pressure. This pressure reduction usually occurs through isenthalpic processes in choke-type valves or in separator and treater vessels.

Energy recovery from oil flow at high pressure and temperature is an excellent energy efficiency alternative. The recovery of the thermal energy of the produced flow is quite common, since, at the very least, it is not wasted along the existing processes in the installation; however, the pressure potential energy is degraded along the productive process. This occurs due to the challenging and ill-behaved character of the multiphase flow, in addition to the way it will expand being little predictable and the presence of solid and corrosive impurities in it. For this source to be better used, multiphase expanders must be developed and they must be robust enough to handle the multiphase flow, in addition to the need to use materials resistant to the presence of solids and corrosion. Currently, in addition to the limitation related to the commercial existence of this type of equipment, there is obviously an economic difficulty due to the need to build the machine with high-cost noble materials.

The potential for this recovery or use of energy depends on the thermodynamic properties at the entrance to the installation, that is, it depends on the temperature and pressure levels of the flow.

At this point, we take the liberty of inviting the reader to carry out a somewhat philosophical reflection:

Nature took millions of years to generate and store hydrocarbons in underground reservoirs. Throughout this time, layers of sediments were deposited in order to compress these reserves, a fact that significantly increases the pressure and temperature of the stored hydrocarbons. Would it be fair then, in a matter of a few minutes, to degrade the pressure of this flow and then compress it again for transport, with the contribution of additional energy resources and impact on the environment?

5.7.5 Heat Transfer Enhancement

There is not always a need for optimization of thermal exchange processes in production facilities, due to the fact that there are excessive thermal sources in production processes, especially when we are referring to the production of light oil, which demand low thermal energy in the stabilization process. However, when it comes to the production of heavy oil, every effort must be made to avoid the need for thermal sources in addition to those commonly existing in production facilities, such as the use of furnaces. In this way, the increase in the efficiency of thermal exchanges can collaborate intensely with this commitment.

The development of compact exchangers with a high heat exchange surface using conventional technology will be one of the possible solutions to be used. However, issues related to fluid cleanliness will be a challenge in this development.

An alternative that could be used is the development of solutions using nanotechnology, with the emergence and injection of nanofluid additives with the property of intensifying heat transfer and/or with the use of nanosurfaces in the construction of heat exchanger exchange areas in order to increase the heat exchange surface.

5.7.6 Use of Renewable Energy Sources

In the future, with the reduction in the cost of renewable energies, the use of these sources as a secondary form of energy conversion will gradually become more common. In onshore production facilities, the use of solar energy for simple heating of the produced fluids to be stabilized, for steam generation (in production of heavier oils) and for electric energy generation in photovoltaic panels or in Rankine cycles will gradually be more common.

In offshore installations, the use of wind energy, as well as wave energy and ocean currents, may be feasible. Solar energy will have a more restricted application due to the large areas needed.

However, the use of renewable energy sources will only be considered an option if the pressure to reduce atmospheric emissions of pollutants and greenhouse gases continues to increase and the investment costs in these sources become competitive compared to natural gas produced.

5.8   Theoretical Reference of Exergy Efficiency of an Oil Production Facility

Figure 5.18 presents the theoretical exergy flow in a gas turbine (typical primary energy conversion process of a production facility). Only a fraction of 35% of the chemical exergy of the fuel is converted into mechanical work, another 30% is released by the exhaust and would have potential for useful use, if a cycle closure were applied (combined cycle or cogeneration). Thus, this primary source would be limited to a maximum exergy efficiency, around 65%. We invite the reader to revisit Chap. 2 on Exergy.

The difference between 100 and 65% is due to the primary irreversibilities in the gas turbine, mainly exergy destruction (entropy generation) in combustion.

The presentation of exergy efficiency values with small magnitudes, in absolute terms, can lead to a misunderstanding, mainly by the lay public in thermodynamics. For this reason, it becomes necessary to establish a maximum level to be reached, demonstrating that the reference is no longer 100%, but a value below this according to the source or primary conversion of energy of the production plant, as discussed in paragraphs above.

Thus, a production facility that has gas turbines as its primary energy sources presents exergy efficiency limited by this inefficient form, already in its initial base of conversion of the chemical exergy of the fuel gas into electrical or mechanical work.

As the difference between 100 and 65% occurs due to primary irreversibilities in the gas turbine, mainly destruction of exergy (entropy generation) in combustion, we could only reverse this situation if there was a change in primary technology or technological improvement.

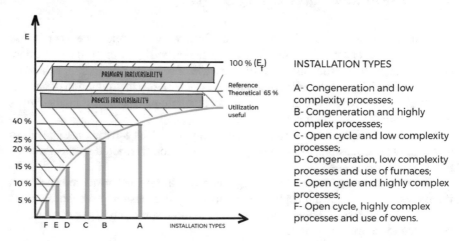

Fig. 5.18   Types of installation in relation to theoretical exergy efficiency

If all the other processes in the unit were perfect, that is, reversible, and there was a thermal demand for all the thermal energy in the exhaust of the machines, the maximum theoretical exergy efficiency of the plant would be 65%. However, there are other processes and they are real and therefore irreversible. Therefore, the exergy efficiency of a system, whose primary source is a cogeneration system, or a cycle combined with a gas turbine and in addition to this source has other processes, will never have an exergy efficiency of 65%. The more energy conversion processes that exist and/or the more these processes are irreversible, the farther the system will be from the theoretical reference. Furthermore, in most of the projects and existing stationary production units, there is no thermal demand for all the energy contained in the exhaust of these combustion machines. As the installation of a combined cycle in a marine unit is something rare, the exergy efficiency of these plants is naturally low.

Figure 5.16 demonstrates the growth of exergy efficiency, from zero to asymptotic approximation to the reference. The zero value of efficiency occurs in a system that allows the primary source to come into balance with the environment without any useful benefit. As the system uses the exergy of the fuel in a useful way, preferentially transferring it to the products, the exergy efficiency grows towards the theoretical reference value up to a maximum level, defined as a function of the number of processes in subsequent conversions (complexity of the plant), level of irreversibilities of these processes and existence of thermal demand in the cogeneration installation. In other words, the unit will never reach the maximum theoretical reference, because not only is the primary conversion irreversible (Primary Irreversibilities), but there are other processes, which are also irreversible and therefore destroy exergy (Process Irreversibilities). The asymptotic behavior of the curve shown in the figure is due to the existence, in addition to the primary energy conversion, of one or more real processes that convert energy. This figure also identifies the theoretical decrease of the exergy efficiency according to the increase in the complexity of the processes and the reduction in the use of thermal energy from the exhaust of the turbines in the installations or projects of oil production units. (The values presented in the figure are not calculated, they are only estimated based on the qualitatively evaluated results.)

In terrestrial installations fed with external electrical energy, the exergy efficiency, exclusively of the installation, is naturally higher, since the primary irreversibilities responsible for a very considerable portion of the inefficiencies occur in the primary energy conversion process, which, in this case is external to the evaluated control volume.

References

Rodrigues, P. S. B. (1991). Compressores Industriais. Rio de Janeiro, Editora Didática e Científica Ltda.

Szargut, J., Morris, D. R., & Steward, F. R. (1988). Exergy analysis of thermal, chemical, and metallurgical processes. New York/Berlin, Hemispere/Springer.

Chapter 6
Challenges and Opportunities for Oil in the Context of Deep Decarbonization

The shift to cleaner forms of energy must happen as quickly as possible in order to reduce emissions of greenhouse gases. However, the viability of the available possibilities in terms of energy is something that needs to be analyzed so that the studies may be carried out based on trajectories that are able to meet human requirements in all their guises. Thus, the purpose of this study is to identify the opportunities and problems associated with the transition to a cleaner energy source, with a particular focus on the oil industry. These opportunities and challenges are broken down according to the environmental, economic, technological, and social dimensions. In particular, the paper titled "Net Zero by 2050 A Roadmap for the Global Energy Sector" that was published by the International Energy Agency (IEA) is evaluated. The findings suggest that oil will continue to play an important role, mainly in the transportation business, the petrochemical industry, and fundamental industry; nevertheless, adjustments to the low carbon scenario will be required. The significance of businesses in the oil and gas industry is also brought to light. Act in conjunction with national governments to increase energy access and the exchange of technological information while simultaneously lowering overall usage.

In light of this, we discuss the potential and difficulties that lie ahead for the oil industry as a result of the energy shift. The technological and environmental components both offer chances for the adoption of solutions that reduce greenhouse gas emissions. Some examples of these technologies include carbon sequestration and storage as well as the utilization of renewable energy sources.

Nevertheless, there will be a lot of obstacles to overcome, as carbon sequestration technologies are not quite where they need to be yet, and emissions will need to be cut down at every stage of the production chain. The diversity of energy sources, such as natural gas and renewable energies, can assist businesses in maintaining their relevance in a market that is undergoing transition from an economic and governance perspective. Nevertheless, the energy transition may provide a risk to the conventional economic model of the oil industry, and the push from investors and civil society for more environmentally responsible practices may present a difficulty for the corporate

M. V. da Silva Neves and A. F. Flutt, *Energy Efficiency in Oil Production*,
SpringerBriefs in Applied Sciences and Technology,
https://doi.org/10.1007/978-3-031-54274-9_6

governance of corporations operating in the area. According to the IEA research that was referenced, there should be a major fall in the demand for oil and gas, with the use of fossil fuels being limited to the production of commodities in which carbon is an integral part of the product and to industries in which there are few possibilities for the reduction of emissions. The electrification of transport is an important strategy for lowering emissions. The use of fossil fuels should be discouraged by public policy, and new investments in oil and gas fields should be restricted, with the goal of concentrating supplies on producers with the lowest costs. In order to compensate for the declining profits from oil and gas, structural reforms and new sources of revenue are required.

6.1   Energy Return on Investment and the Competitiveness of Oil

From the perspective of EROI, the current energy transition faces the challenge of intentionally replacing higher EROI sources with lower EROI sources, which is precisely the opposite trend of previous energy revolutions (Smil, 2011).

Petroleum currently leads the pack in EROI evaluations, with its high energy density and relative ease of extraction contributing to a significantly higher EROI than most renewable energy sources (Neves et al., 2022). As Hall et al. (2014) noted, petroleum's EROI ranges between 15:1 and 20:1, while solar and wind power typically fall between 5:1 and 10:1, as presented in Fig. 6.1. This disparity means that for every unit of energy invested in oil production, we gain 15–20 units in return, as opposed to only 5–10 units from renewable sources.

The superior EROI of petroleum propels its competitive edge over most renewable energy sources. It is a highly concentrated energy source, cost-effective in extraction and transport, making it an attractive choice for various energy applications, from transportation to electricity generation.

Nonetheless, it's vital to recognize that the EROI of oil isn't static. As petroleum reserves deplete, extracting the remaining reserves becomes more challenging and energy-intensive, consequently decreasing its EROI over time and potentially undermining its competitiveness against renewable sources.

Furthermore, we cannot overlook the environmental impact of oil extraction and combustion. Despite its EROI advantages, petroleum's negative externalities, including air and water pollution, greenhouse gas emissions, and ecosystem degradation, are substantial. Renewable energy sources, in contrast, have much lower environmental impacts, marking them as more sustainable and desirable options in the long run.

While the EROI is a valuable metric, it doesn't capture every aspect of the current energy transition, particularly the urgent need to reduce CO_2 emissions. Cleveland

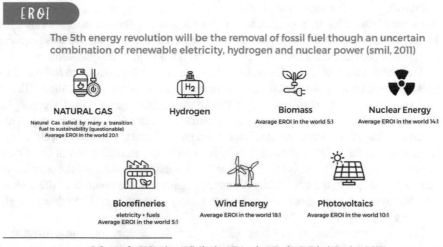

Fig. 6.1 Sources predicted by Smil (2011) for the energy transition

(1984) explains that the original application of EROI aimed at finding energy alternatives that could sustain economic growth while matching or exceeding oil's efficiency. Today's challenge lies in introducing energy alternatives that mitigate CO_2 emissions, despite their typically lower EROI (Hall et al., 2017). Thus, the selection of technologies that balance efficiency, economic return, and environmental impact mitigation must consider both EROI and CO_2 emissions.

Moreover, the traditional EROI evaluation approach, relying on calorific values to quantify direct and indirect energy input and output, has limitations. It only estimates the quantity of energy, disregarding energy quality, a crucial factor determining an energy source's societal utility. It also fails to provide a comprehensive view of the system's complexity, like labor, auxiliary services, and environmental inputs. To address these issues, physical approaches such as emergy and exergy analysis have been developed. These account for both the quality and complexity of the system, and other physical approaches include ExROI, Return of Exergy on Investment in Exergy, and minimal exergy return rates required by society.

In this context, there is no "silver bullet" that can decarbonize the economy; rather, what is required is a systematic worldwide coordination of agents working towards a common objective (Lofgren & Roótzen, 2021). There is a degree of inertia associated with this change brought on by the current economic and energy framework, which is optimized for operation using fossil fuels. Because it is difficult to replace fossil fuels in the transportation sector, steelworks, petrochemicals, and the cement sector, the focus in these sectors in the short term should be on increasing the energy efficiency of processes and applying Carbon Capture, Utilization and Storage (CCUS) technologies that are already available, while massive investments are made in research and development for alternative fuels for the long term. Specifically, petrochemicals show themselves as a prospective niche for moving the demand for oil, with the advantage of embedding carbon in the product that would be used to produce transition infrastructure. This could be a potential solution to the problem.

The mobilization of companies in the oil and gas sector to support the reduction of emissions that do not generate an increase in social well-being—for example, illegal deforestation—may allow for a longer use of petroleum derivatives, for example, allowing for increased safety energy in countries that are lacking in energy. This is because illegal deforestation does not generate an increase in social well-being. In addition, industrialized nations will have to take the initiative to cut their own consumption in order to designate a portion of their carbon budget for the purpose of narrowing the gap in the availability of energy sources. In the absence of this reduction, achieving climate goals and reducing energy poverty may become mutually exclusive objectives.

In conclusion, although oil's high EROI grants it considerable competitiveness over most renewable sources, we must consider the long-term sustainability and environmental impacts of our energy choices. As we transition towards a sustainable energy future, we need to continue investing in renewable energy technologies and mitigating our current energy systems' negative impacts. The challenge of transitioning from high EROI sources to lower EROI sources demands careful planning and infrastructure investment to effectively integrate renewable sources into our energy system.

References

Cleveland, C., Costanza, R., Hall, C., & Kaufmann, R. (1984). Energy and the US Economy: A biophysical perspective. *Science, 225*(4665), 890–897.

Hall, C. A. S., Lambert, J. G., & Balogh, S. B. (2014). EROI of different fuels and the implications for society. *Energy Policy, 64*, 141–152. https://doi.org/10.1016/j.enpol.2013.05.049

Löfgren, Å., & Rootzen, J. (2021). Brick by brick: Governing industry decarbonization in the face of uncertainty and risk. *Environmental Innovation and Societal Transitions, 40*, 189–202. https://www.sciencedirect.com/science/article/pii/S2210422421000435

Smil, V. (2011). Science, energy, ethics, and civilization. In *Visions of discovery: new light on physics, cosmology, and consciousness* (pp. 709–729). Cambridge University Press. https://books.google.com.br/books?hl=pt-BR&lr=&id=BhcpiZN2MOIC&oi=fnd&pg=PR7&dq=SMIL,+Vaclav.+2010.+Science,+energy,+ethics,+and+civilization.+Cambridge:+Visions+of+Discovery:+New+Light+on+Physics,+Cosmology,+and+Consciousness,+R.Y.+Chiao+et+al.+eds.,+pp.+709-729+-+Cambridge+University+Press.&ots=GXdAwNtzzh&sig=3VJVb1XUG_dzgtaQ5Q9dZZwOoXc#v=onepage&q&f=false

Uncited References

Bejan, A. (2016). *Advanced engineering thermodynamics* (4th ed.). Wiley.

Boles, M. A., Yunus A, & Cengel, D. (2014). *Thermodynamics: An engineering approach*. McGraw-Hill Education.

Hall, C. A. S. (2017). *Energy return on investment: A unifying principle for biology, economics and development*. Springer. https://www.researchgate.net/publication/312008285_Energy_Return_on_Investment

Neves, M. V. S (2008) *Exergy efficiency of stationary oil production units, COPPE/UFRJ* [M.Sc. Dissertation]. Rio de Janeiro, Brazil, XVI, 152p, 2008.

Neves, M. V. S., Aylmer, R. R. B. A., & Szklo, A. (2022). Energy transition: Opportunities and challenges for oil in the context of deep decarbonization. *Rio Oil and Gas Expo and Conference*, 22(2022), 195–196. https://doi.org/10.48072/2525-7579.rog.2022.195

Sanjuán, M. A. F. (2011). Concepts in thermal physics, 2nd ed., by Stephen J. Blundell and Katherine M. Blundell. *Contemporary Physics*, 52(1), 98–98. https://doi.org/10.1080/00107514.2010.529508

Smil, V. (2019). Energy (r)evolutions take time. *World Energy, 44*, 10–14.

Printed in the United States
by Baker & Taylor Publisher Services